Distributed Secondary Control of Microgrid Systems

This book presents a detailed description of the transition from the traditional power system to the microgrid (MG) system.

The authors introduce the basic concepts of an MG system along with its different operating modes and classifications. The various benefits, challenges, and technical aspects of an MG are highlighted. They demonstrate different control strategies that are applied at different levels of the control hierarchy, particularly the distributed secondary control architecture. Furthermore, the adaptive techniques-based distributed schemes, a distributed architecture-based synchronization controller, and a delay-independent buffer-free distributed controller are proposed for the MG system.

This book can be useful for senior undergraduate students, postgraduate students, and researchers of electrical power engineering and system control.

Haoming Liu received the B.S., M.S., and Ph.D. degrees all from Nanjing University of Science and Technology, China, in 1998, 2001, and 2003, respectively. He is currently a full professor at the School of Electrical and Power Engineering, Hohai University, China. He has published more than 200 journal papers, 30 patents, and two books. His research interests include new energy generation, distribution system analysis and control, and the electricity market.

Muhammad Yasir Ali Khan received his Ph.D. degree from Hohai University in 2023. He is now a postdoctoral researcher at the College of Water Conservancy and Hydropower Engineering, Hohai University, China. His research interests include power electronic converters, renewable integration, microgrids, and power systems.

Jingjing Zhai received her Ph.D. in Control Engineering from Nanjing University of Science and Technology, China, in 2022. She is currently an associate professor at Nanjing Institute of Technology. Her research interests include the optimal operation of integrated power systems and intelligent power utilization.

Distributed Secondary Control of Microgrid Systems

Haoming Liu, Muhammad Yasir Ali Khan,
and Jingjing Zhai

CRC Press
Taylor & Francis Group
Boca Raton London New York

CRC Press is an imprint of the
Taylor & Francis Group, an **informa** business

Designed cover image: © Haoming Liu

MATLAB® and Simulink® are trademarks of The MathWorks, Inc. and are used with permission. The MathWorks does not warrant the accuracy of the text or exercises in this book. This book's use or discussion of MATLAB® or Simulink® software or related products does not constitute endorsement or sponsorship by The MathWorks of a particular pedagogical approach or particular use of the MATLAB® and Simulink® software.

First edition published 2025
by CRC Press
2385 NW Executive Center Drive, Suite 320, Boca Raton FL 33431

and by CRC Press
4 Park Square, Milton Park, Abingdon, Oxon, OX14 4RN

CRC Press is an imprint of Taylor & Francis Group, LLC

© 2025 Haoming Liu, Muhammad Yasir Ali Khan and Jingjing Zhai

ISBN: 978-1-032-97554-2 (hbk)
ISBN: 978-1-032-97555-9 (pbk)
ISBN: 978-1-003-59428-4 (ebk)

DOI: 10.1201/9781003594284

Typeset in Minion
by codeMantra

Contents

Acknowledgments

W<small>E WOULD LIKE TO</small> express our heartfelt gratitude to all those who have contributed to the creation of this book.

First and foremost, we extend our deepest appreciation to Jian Wang from the School of Electrical and Power Engineering, Hohai University and Danish Khan from Hong Kong Industrial Artificial Intelligence & Robotics Centre for their unwavering support, insightful guidance, and encouragement throughout this project. Their expertise and thoughtful feedback have been invaluable in shaping our ideas and refining our arguments.

We would also like to thank our colleagues and friends at Zhihao Yang, College of Electrical, Energy and Power Engineering, Yangzhou University; Tasawar Abbas, School of Electrical and Power Engineering, Hohai University; Zhixin Fu, School of Electrical and Power Engineering, Hohai University; and Yue Pu, School of Electrical and Power Engineering, Hohai University for their stimulating discussions and camaraderie. Your insights and collaboration have enriched our work and provided a supportive environment in which to explore new ideas.

A special acknowledgement goes to our families for their enduring support and belief in our work. Their encouragement has been a source of strength, motivating us to persevere through challenges.

Thank you all for your invaluable contributions to this journey.

List of Abbreviations

API	Adaptive Proportional Integral
BP	Back Propagation
BSS	Battery Storage System
CAN	Customer Premises Area Network
CN	Communication Network
CT	Communication Technology
CTDs	Communication Time Delays
DAPI	Distributed Averaging Proportional Integral
DG	Distributed Generated
DNO	Distribution Network Operator
DR	Demand Response
DSL	Digital Subscriber Line
DTR	Data Transfer Rate
EMS	Energy Management System
ESS	Energy Storage System
FACTS	Flexible AC Transmission System
FLC	Fuzzy Logic Controller
FPI	Fuzzy Proportional Integral
FR	Fuzzy Rule
GC	Grid Connected
GF-DG	Grid Forming DG
GHG	Green House Gases
GMF	Gaussian Membership Function
HDI	Human Development Index
HMG	Hybrid Microgrid
HN	Hybrid Network
HV	High Voltage
IS	Islanded
LoWPAN	Low-Powered Wireless Personal Area Network

LV	Low Voltage
MAS	Multi-Agent System
MG	Microgrid
MGCC	Microgrid Central Controller
MIT	Massachusetts Institute of Technology
mMTC	Massive Machine Type Communication
MPC	Model Predictive Controller
MPPT	Maximum Power Point Tracking
MV	Medium Voltage
NAN	Neighborhood Area Network
PCC	Point of Common Coupling
PE	Power Electronics
PLC	Power Line Communication
PLL	Phase-Locked Loop
PnP	Plug and Play
PR	Proportional Resonant
P&O	Perturb and Observe
PV	Photovoltaic
PWM	Pulse Width Modulation
R-DG	Regulating-DG
RESs	Renewable Energy Resources
SC	Secondary Control
SCADA	Supervisory Control and Data Acquisition
SD	Steepest Descent
SG	Smart Grid
SMC	Sliding Mode Controller
STATCOM	Static Synchronous Compensator
SYNC	Synchronous Controller
UG	Utility Grid
ULLRC	Ultra-Low Latency and Reliable Communication
VOI	Virtual Output Impedance
VP	Voltage Phasor
VPP	Virtual Power Plant
VSI	Voltage Source Inverter
WAN	Wide Area Network
WDCTs	Wired Communication Technologies
WSCTs	Wireless Communication Technologies

CHAPTER 1

An Overview of Microgrid Concept, Classifications, and Components

1.1 TRANSITION FROM TRADITIONAL DISTRIBUTION NETWORK TO SMART GRID

A power system is composed of interconnected components that generate, transmit, and distribute electrical power to end-users. These components are linked together using transformers, which are responsible for adjusting the voltage to a level which is suitable for the proper functionality of the system.

In the early 20th century, the organization and structure of the electrical grid underwent extremely slow growth. Only fossil fuels were used for energy generation, and the generating units were located far away from the load centres. Moreover, the transmission networks were organized in a mesh structure, while the distribution networks were organized in a radial structure to provide electricity to the end-users. A generalized structure of traditional power systems comprised of generation, transmission, distribution, and end-users, interconnected through substations and power lines is presented in Figure 1.1.

DOI: 10.1201/9781003594284-1

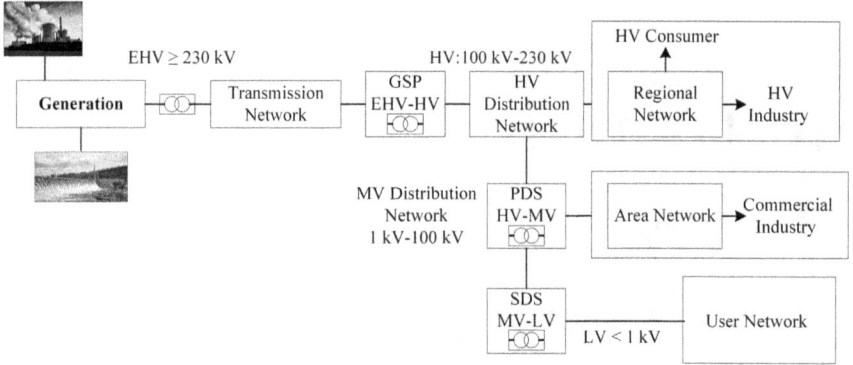

FIGURE 1.1   Structure of conventional power system.

From Figure 1.1, it can be seen that in a traditional system, a large-scale centralized power generating unit is used to generate the electrical power. A generating power is then injected into the high voltage (HV) transmission network (such as 230 kV and 400 kV) through a step-up transformer. A transmission network then delivers the electrical power to the distribution network through grid supply points that step down the voltage to the distribution HV level, which is generally 100–230 kV. In the distribution network, at a primary distribution substation, a voltage is step down to medium voltage (MV) level ranging from 1 to 100 kV [1]. After that, at a secondary distribution substation, the voltage is further step down to low voltage (LV) levels to make it feasible for the end-users [2]. Finally, through the LV distribution network, the electricity reaches single-phase or three-phase end-users (230 V single-phase or 400 V three-phase) [3].

Depending on the voltage level and load density of the system, three different distribution network topologies are used to supply electrical power to the end-users. These are (a) radial, (b) ring or weakly meshed, and (c) meshed, as presented in Figure 1.2. In a radial distribution network, a substation is connected to each node through a single path. Generally, a ring topology is used in LV networks and in MV distribution networks where isolated load areas are needed to connect with the system through long rural lines. A ring or weakly meshed distribution network uses a radial operation to improve the security of electrical supply in case of scheduled or circuit outages. From Figure 1.2b, it can be seen that a normal open point is located between feeder A and feeder B to ensure the radial operation. Hence, in case of any fault, the location of the normal open point can be removed to ensure a smooth electricity supply to the end-users.

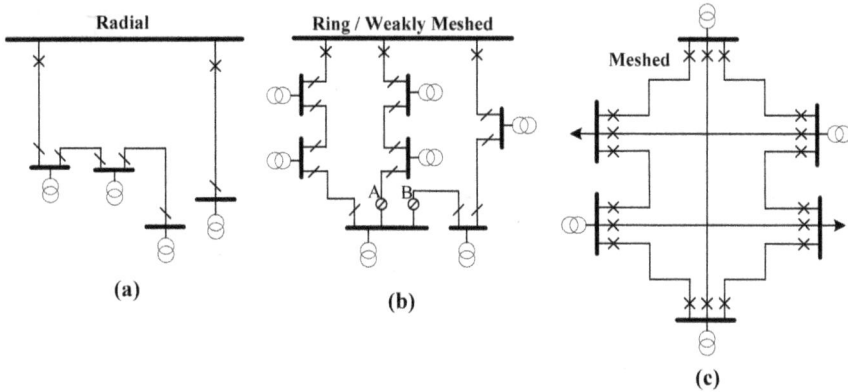

FIGURE 1.2 Distribution network topologies.

Moreover, in some scenarios, when distribution feeders serve high-density load areas, especially in urban areas; hence, in these scenarios, a system may contain several loops, which can be created by closing the normally open point switches. A meshed network is used at the HV level and is applied in large distribution areas such as regional networks. Moreover, the operation of meshed network, as shown in Figure 1.2c, is managed by a transmission or distribution system operator. In a meshed distribution network, multiple substations are connected in parallel, which reduces the number of transformers in a group and increases the security to a greater extent [3]. However, in case of high-voltage system, if any fault occurs, then the parallel operation of laced points results in reverse power flow through the in-feed transformers. Therefore, care should be taken and protocols must be followed such that the level of the faults within the network is in acceptable range.

In traditional power distribution systems, the energy is generated in a central power plant; however, in the last few decades, due to the rapidly varying nature of end-user loads and the high penetration of distributed RESs, a concept of smart grid (SG) has been developed. In an SG, there is no central power generation centre; however, it comprises distributed generation, as presented in Figure 1.3. A traditional power grid transmits power from central generators to many consumers, while in SG, two-way communication occurs that enables a consumer to generate energy and deliver that energy to the grid. SG enhances the power system by providing efficient power and a self-healing feature for many events [4]. The main advantages of SG include improved capacity, increased

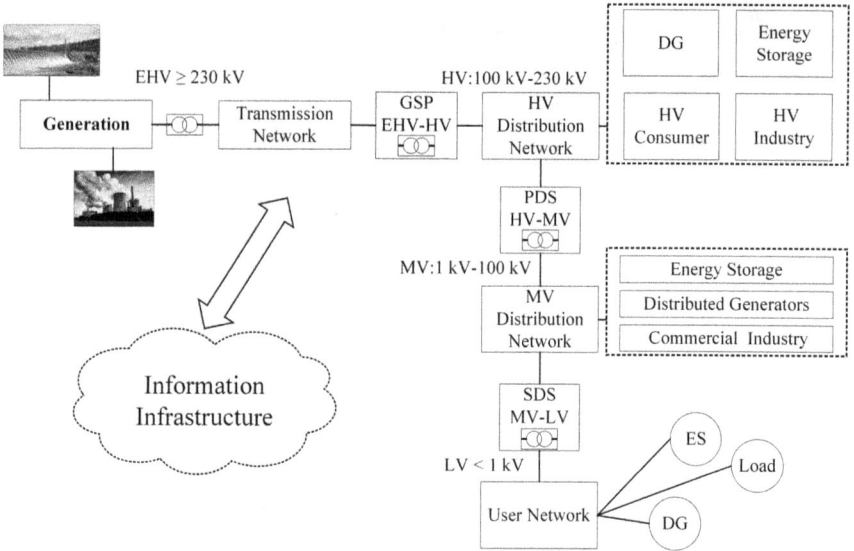

FIGURE 1.3   Transition from traditional power system to SG concept.

TABLE 1.1   Comparison of Traditional Grid and Smart Grid

Traditional Grid	Smart Grid
One-way communication	Two-way communication
Electromechanical	Digital
Few sensors	Sensors throughout
Centralized generation	Distributed generation
Manual restoration	Self-healing
Failure and blackouts	Adaptive and islanding
Manual monitoring	Self-monitoring
Few consumer choices	Many consumer choices
Limited control	Pervasive control

efficiency of power systems, easy maintenance, distributed power generation, lower greenhouse gas emissions, self-healing to disturbances, reduction in oil consumption, more options for consumers, remote monitoring of events, and plug-in electric vehicle technology management [5,6]. A brief comparison of the traditional grid and SG is presented in Table 1.1. Besides these advantages, some technical challenges arise when a new energy-generating source is integrated into an existing distribution system, such as overvoltage, inversion of power flow, stability problems, and many others [7–9].

1.2 FUNDAMENTAL OF MG

The integration of RESs in power networks has increased globally due to zero-emission, universal availability, and low operational cost [10]. Therefore, the world has adopted more and more RES for energy generation. Globally, at the end of 2022, the total installed capacity of RESs is about 3494 GW, in which a share of hydropower is 1210 GW, PV is 1185 GW, wind is 923 GW, bio-power is 151 GW, geothermal is 15.5 GW, concentrating solar thermal power is 6 GW, and ocean power is 0.5 GW. According to the Renewables 2023 (REN21) global status report, globally, for the last six consecutive years, the installation of RES is more than the combined use of fossil fuels and nuclear power. The yearly installed capacity from 2014 to 2022 of different RES is presented in Figure 1.4. From Figure 1.4, it can be seen that in 2022 alone, the total addition of RES is 348 GW, of which a majority is from PV, i.e., 243 GW, which is then followed by wind, i.e., 78 GW, etc. [11].

This fast expansion and growth in the installation of RESs are mainly due to the rising demand for electricity due to population growth and industrial development, the desire to control the energy generation from fossil fuels, an increase in the emission of harmful gases, government support, reduction in component prices, technological improvements, and advancements in RESs integration technologies [12]. However, compared to the large-scale installation, the RESs are generally installed in a distributed manner, close to end-users. Deploying distributed RESs can

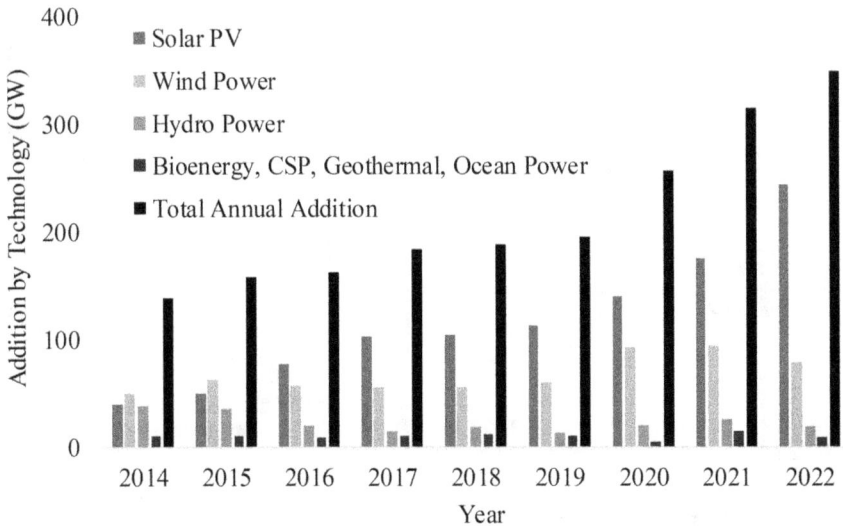

FIGURE 1.4   RES addition by year (2014–2022).

reduce transmission line losses, increase energy efficiency, increase grid resilience, avoid generation costs, and reduce carbon pollution. These new interests lead to various emerging challenges in the power system. Firstly, the power generated from RESs is stochastic and intermittent in nature. Secondly, their availability mostly depends on the geography of the area and environmental conditions. Thirdly, the penetration of a vast amount of RESs in the existing power system has affected the power quality to a great extent due to (a) load frequency drifts, (b) severe outages in case of weather shocks, (c) low voltage generation, etc. Their randomness and intermittent nature may further cause phase and frequency deviation, voltage fluctuations, increased grid uncontrollability, high probability of off-grid, and various safety risks to grid operation.

Hence, to handle the above-discussed problems and uncertainties in a small power system, the concept of SG is shifted towards an MG. An MG as a major component in an SG system can be defined as an electrically bounded area of the low voltage distribution network that aggregates locally distributed generated (DG) units, energy storage elements, and controllable loads to form a self-sufficient energy system and has the capability to operate either in grid connected (GC) or islanded (IS) mode [13]. A generalized structure of an MG connected with the utility grid (UG) through point of common coupling (PCC) is presented in Figure 1.5.

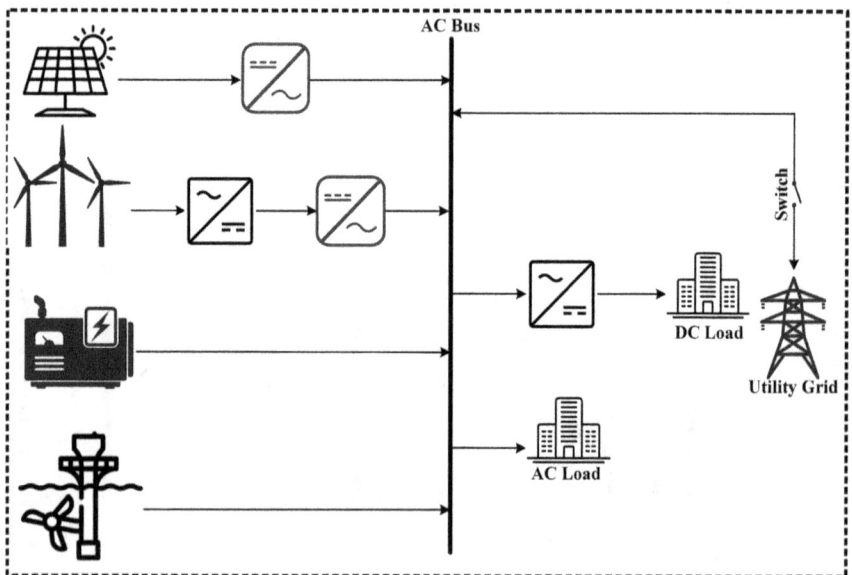

FIGURE 1.5   Generalized conceptual structure of an MG.

It is important to keep in mind that a set of interlinked generators and loads does not necessarily have to be intelligent enough to qualify to be called an MG. Hence, an MG is a structure that is either connected to or isolated from the UG, where load demands are fulfilled by local DG units. Moreover, a high level of intelligence can be applied to some or all the components of MG, such as load, energy management and energy generation, to efficiently optimize its operation and ensure an uninterruptable power supply to its end-users with high energy efficiency and cost-effectiveness [14].

1.3 OPERATING MODES OF MG

An MG has to perform its operation either in IS mode or in GC mode. To achieve smooth performance in these operating modes and ensure a seamless transition of GC mode into IS mode or vice versa, suitable control systems and equipment are required to provide uninterruptible electricity to end consumers. These operating modes are discussed below in detail.

1.3.1 Islanded Mode

During IS mode of operation, an MG works autonomously from the existing power system while maintaining high power quality and an uninterruptible power supply for its consumers. The IS operation of an MG can be intentional or unintentional [15]. Intentional islanding occurs when the quality of the power system network may jeopardize the operation of an MG or during scheduled maintenance. On the contrary, unintentional IS occurs during grid disturbances or unscheduled events. In IS mode, an MG does not receive any support from the UG; therefore, the controller of the MG should be robust enough to regulate the frequencies and voltages of the DGs while attaining appropriate power sharing. Moreover, an MG should be provided with intelligent controllers that ensure a seamless and smooth transition between different operating modes and maintain a continuous power supply to the end-users [16,17].

1.3.2 Grid Connected Mode

When an MG is connected to the UG through PCC, then an MG is said to be in GC mode. In GC mode, the whole performance and operation of the MG is controlled directly by the UG, which also determines the frequency and voltage. During GC mode, the deficit power of the MG must be supported by the UG to meet the load demand. Also, the extra power of the MG must be supplied to the UG to support the ancillary services

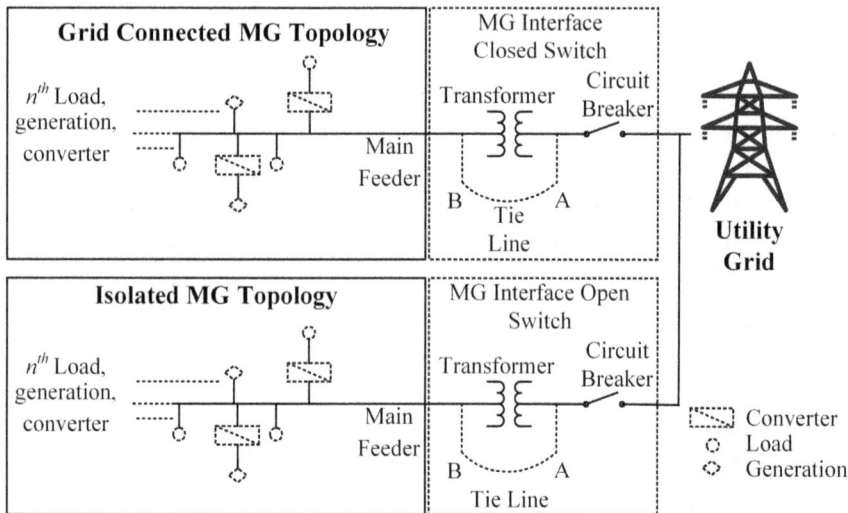

FIGURE 1.6   Operating modes of MG.

such as reserve power and voltage regulation. During this operating mode, the import/export of power is maintained by MG, and MG continuous to operate in this mode until intentional or unintentional IS occurs. The two operating modes of the MG are presented in Figure 1.6.

1.4 CLASSIFICATION OF MGs

MGs can be categorized into seven different groups according to their configuration, application, location, scenario, generating source, control, and size, as presented in Figure 1.7. These categorizations of MG are discussed below in detail.

1.4.1 Classification of MG Based on Configuration

Based on the voltage characteristics and system architecture, the MGs are classified into four categories: (a) DC MG, (b) AC MG, (c) hybrid AC/DC MG, and (d) networked MG. These different types of MGs are discussed in detail below.

1.4.1.1 DC Microgrid

In a DC MG architecture, the main bus is DC; therefore, DC loads can be integrated directly with the bus without using any power electronics (PE) conversion process. However, for AC loads, DC-AC inverters are needed for the inversion to make it applicable for the AC loads. A generalized

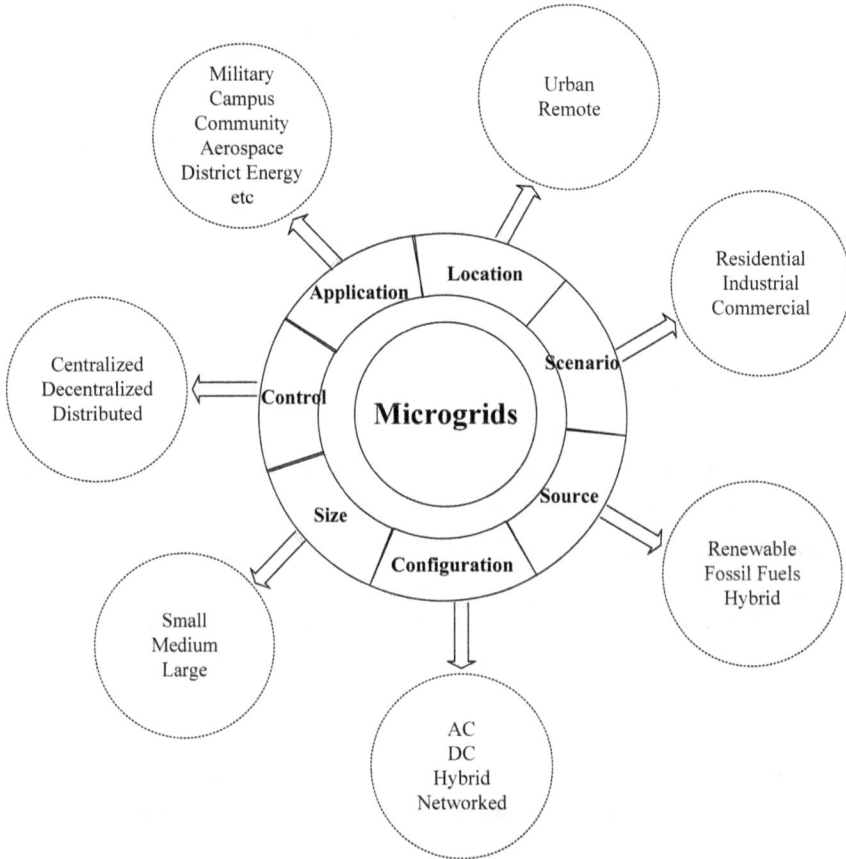

FIGURE 1.7 Classification of MGs.

schematic of the DC MG topology considering different AC/DC loads and DG units is presented in Figure 1.8. In recent years, the increased penetration of RESs has accelerated research in the area of DC MG architecture. In case of PV generation, DC-DC step-up converters are used to integrate the PV system with the DC bus of an MG. In case of wind power generation, AC-DC converters are used to convert the output power of wind turbines (AC in nature) to make it feasible for DC MG application.

DC MG offers numerous advantages, such as (a) high energy efficiency, (b) low energy conversion losses due to less usage of power electronic converters, (c) DC-based DGs can be integrated and controlled easily, (d) direct interconnection of battery storage system (BSS), (e) RESs generation and load fluctuation can be managed easily by using BSS to complete the power deficiency, (f) it is easier to dump the circulating current between

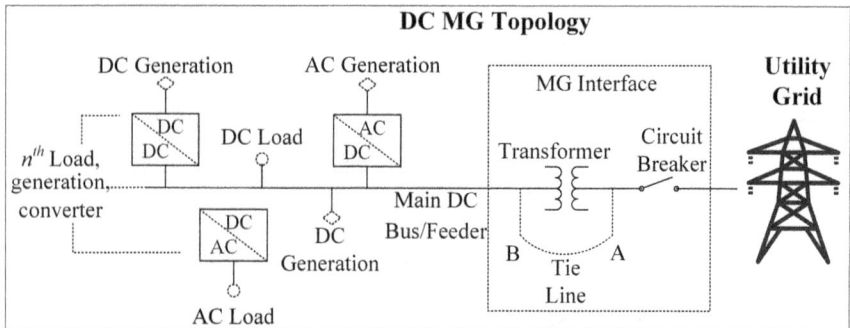

FIGURE 1.8   Generalized structure of DC MG system.

RESs, and (g) simple grid integration [18]. Despite these numerous advantages, there are some disadvantages to the DC MG topology, such as (a) majority of the load is AC, and therefore, DC to AC inversion is required; (b) in a DC distribution system, a voltage drop occurs as the distance from the generation increases, and therefore, voltage boosters are required; and (c) an AC-DC rectifier is required for the AC generating units [19,20].

1.4.1.2 AC Microgrid

In the AC MG configuration, all the generation resources and load are connected to the common main AC bus. Based on the distribution structure, this configuration can be classified into three sub-categories that include three phases with a neutral line, three phases with no neutral line, and a single phase. In this configuration, AC loads can be directly fed without using any PE conversion mechanism. However, for DC loads, AC-DC rectifiers are used. Similarly, for DC power generating units such as PV, DC-AC inverters are used to make them compatible with the AC bus. Moreover, it is simple to integrate the AC MG with the UG as compared to the DC MG [21]. A generalized schematic of the AC MG configuration considering different AC/DC loads and DG units connected with the UG through a circuit breaker is presented in Figure 1.9.

The main advantages of AC MG configuration are (a) they are connected to medium and low voltage distribution networks that enhance the power flow and decrease the power losses in the transmission line, (b) voltage stability, (c) autonomous control of reactive power, (d) highly efficient transformers and (e) simple quenching procedure of fault arc current at zero crossing [22]. Besides these numerous advantages, there are some demerits that are associated with this configuration, such as (a) for DC

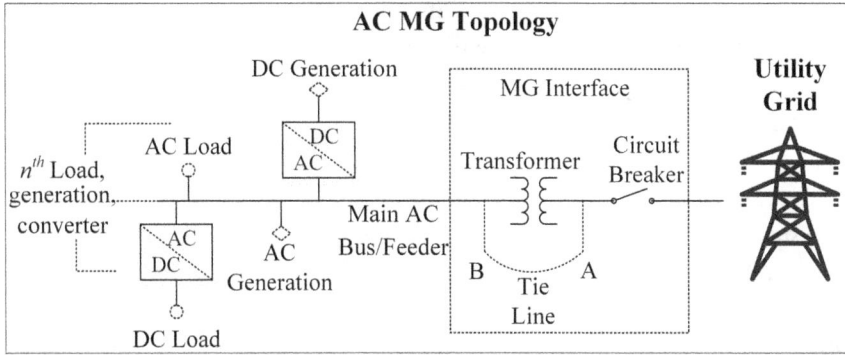

FIGURE 1.9   Generalized structure of AC MG system.

loads, AC-DC converters are needed that greatly affect the system's efficiency; (b) the inclusion of DC-AC inverters for integrating the DC generating DGs; (c) the introduction of harmonics [23]; (d) synchronization
of the voltage phasor (VP) of the AC MG with the UG is required, which
makes the management and control of the AC MG more complex than the
DC MG [24]; (e) synchronization of RESs-based DGs is complex; and (f)
shortage of reactive power in case of RESs-based DGs, especially PV.

1.4.1.3 Hybrid Microgrid

A hybrid microgrid (HMG) configuration consists of both DC and AC
sub-grids, with the main objective being to minimize the number of interface devices and conversion stages while keeping the price of energy low
[25]. Thus, as a result, the overall reliability and efficiency of the system
have improved. In HMG configuration, the power flow between the UG
and the networks is controlled by static switches and PE converters, respectively. A balance between supply and demand determines the direction of
the power flow. To interface both AC and DC MGs, in this configuration,
bidirectional AC-DC interlinking converters are used. Moreover, DC-DC
step-up converters are used for integrating DC-generated energy resources
with the DC sub-bus, while DC-DC step-down converters are used for
DC loads, and bidirectional DC-DC converters are used for integrating
the energy storage system (ESS) with the DC sub-grid. The interlinking
converter between DC and AC sub-grids works according to the overload conditions. When an AC sub-grid is in an overload condition, then
the interlinking converter acts as an inverter, and the direction of power
flow is from DC to AC sub-grid. On the contrary, when the DC sub-grid

is in overload condition, then the direction of power flow is from AC to DC sub-grid, and the converter will act as a rectifier. Hence, in most of the literature, the interlinking converter is considered as the key converter that controls the power flow between the DC and AC sub-grids of the HMG configuration [23]. A generalized structure of HMG is presented in Figure 1.10.

The main advantages of HMG configuration include (a) reduced conversion stages, thus reducing conversion losses; (b) DG units can be tied to the DC or AC bus; (c) reliable ancillary services; (d) Plug and Play (PnP) type management of DC or AC sub-grids; (e) improved load support; (f) UG voltage support; (g) increased economic operation; and (h) overcome

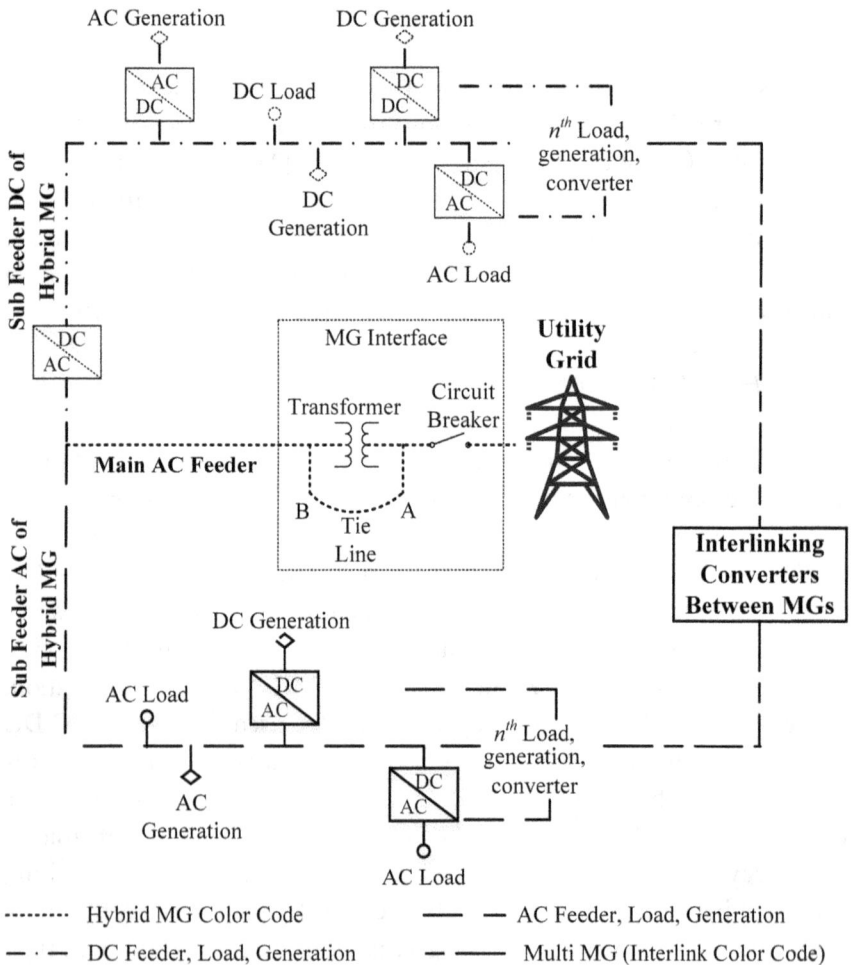

FIGURE 1.10 Generalized structure of hybrid MG system.

the disadvantageous features of individual AC or DC MG. Besides these numerous advantages, the main demerits of this configuration are (a) the inter-relationship between DC and AC sub-grids introduces several operational issues and makes the configuration very complex [23] and (b) less reliable compared with the individual AC MG [26].

1.4.1.4 Networked Microgrid

In a networked MG configuration, numerous MGs are connected with each other to improve the reliability of the system. A generalized schematic of this configuration is presented in Figure 1.11. In this configuration, every MG supports other MGs in case of system disturbance or uncertainty. The main objective of networked MG is to improve reliability, stability, and reduce load shedding in distinct MGs. In multi-MG, different MGs are interconnected into distribution feeders incorporating dynamic or fixed electrical boundaries through a distributed or centralized controller. Different utility companies that adopted multi-MG configuration are the Chattanooga Electric Power Board, Illinois Institute

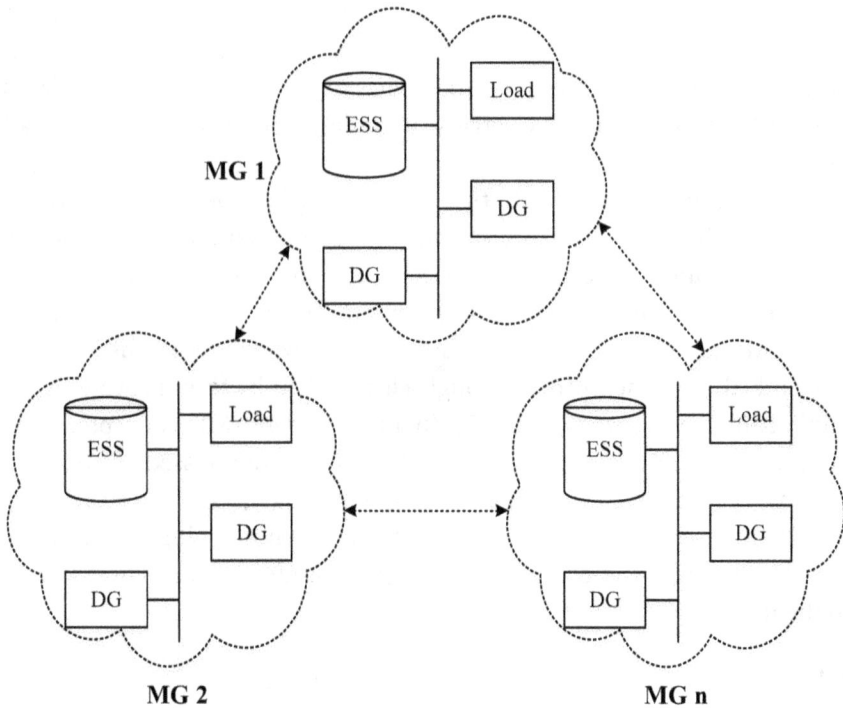

FIGURE 1.11 Generalized schematic of networked microgrid.

of Information Technology Chicago, and Commonwealth Edison [27]. A new concept within multi-MG configuration is the nested MG, which is investigated by the New Paltz MG project [28]. In the nested MG configuration, an MG is divided into nodes that are responsible for ensuring the system stability and providing an uninterruptable power supply to its users within their specified geographic areas. The operation and formation of this method can occur in different ways depending on operational and technical goals.

Based on the electrical boundaries, the nested MG can be classified into two categorizations, i.e., multi-MG with dynamic and fixed electrical boundaries. In state-of-the-art literature, the concept of the dynamic boundary is also stated as grid sectionalization. The boundaries are self-organized and are assisted by the inclusion of new RES-based DG units. The dynamic adjustment of the electrical boundaries of nested MG is possible with the help of voltage and frequency regulation mechanisms. Hence, these fixed boundaries are flexible and operated by using switch gears that may perform as temporary PCCs for the new boundaries. This flexible boundary approach also led to a new concept, i.e., virtual MG. In a fixed electrical boundary, the overall electrical boundary of the system is defined after the merging of different interconnected MGs. The overall boundary is naturally defined by the PCCs and switches. Numerous MGs are clustered within fixed electrical boundaries to ensure that the load demand in balanced.

The concept of virtual MG that is mentioned above is similar to the concept of virtual power plant MG but with additional capabilities such as the incorporation of storage devices, load components, and heterogeneous generation and having the grid-forming ability. Likewise, the concept of nested MG is close to the multi-MG concept, in which numerous MGs are interlinked with each other through electrical links to improve efficacy and facilitate power exchange [27]. In all these multi-MG concepts, there are numerous research challenges that need to be addressed, such as (a) communication problems, (b) control and management of the system, (c) information security, and (d) increasing the system's intelligence in order to ensure stability and provide a reliable and uninterruptible power supply to the users.

1.4.2 Classification of MG Based on Control Structure

According to the control structure, there are three types of MG system, namely, centralized, decentralized, and distributed control structure.

1.4.2.1 Centralized Control Structure

In this control structure, a central controller provides the required control directions to the system to perform its control actions by using a two-way communication network (CN). Due to the high penetration of RESs into a system, it is difficult to employ a centralized control architecture for many reasons, such as limited communication capability among the DGs as well as the limited computation ability of a single controller. Moreover, the extensive usage of a CN makes this architecture more complex, expensive, and unreliable. A generalized schematic of centralized control structure is presented in Figure 1.12a.

1.4.2.2 Decentralized Control Structure

To cope with the limitations of a centralized control structure, a decentralized control structure has been proposed to design a local controller for each DG. For simplicity, each local controller is designed by ignoring the interactions from other DGs and only using its locally available information, as presented in Figure 1.12b. Thus, controllability is restricted by the decentralized approach, and system control performances are deteriorating. Moreover, a lack of communication between the controllers makes the implementation of secondary and tertiary control a challenging task [29].

Large Scale System

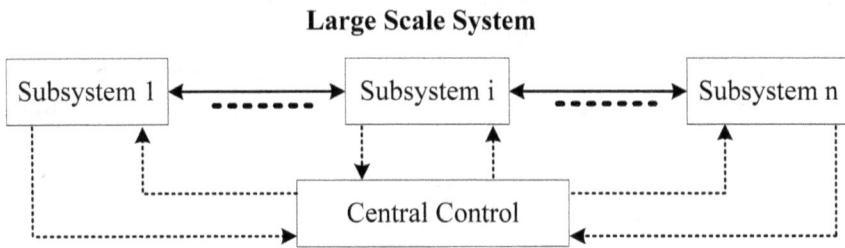

FIGURE 1.12a Schematic of centralized control structures.

Large Scale System

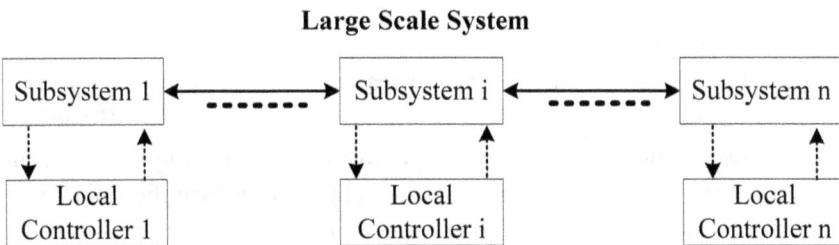

FIGURE 1.12b Schematic of decentralized control structures.

Large Scale System

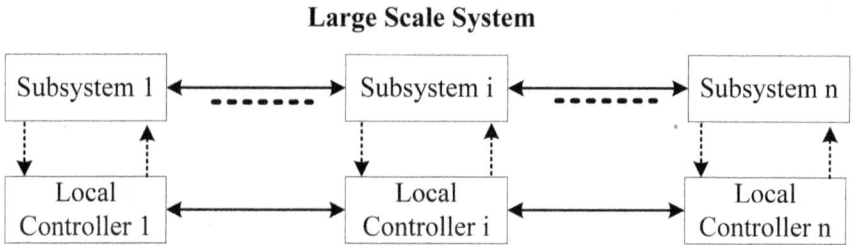

FIGURE 1.12c   Schematic of distributed control structures.

One typical example of the consequences of the drawbacks of such a control strategy is the widespread blackout of August 2003 in North America [30]. In that accident, each DG only focused on maintaining its own subsystem stability and transferred extra load to another DG, which made the overload more severe and eventually caused a blackout.

1.4.2.3 Distributed Control Structure

Distributed control strategies allow information exchange among local controllers by establishing a CN topology among them, as shown in Figure 1.12c. In fact, distributed control strategies can be considered as a trade-off between centralized and decentralized controllers as they combine the advantageous features of both controllers. There are also other factors that motivate distributed control, such as (a) high feasibility for PnP functionality, (b) low cyber cost, (c) low bandwidth CN, and (d) high robustness to single-point failure [31].

1.4.3 Classification of MG Based on Location

One the basis of location, the MGs can be classified into two types: urban MGs and remote MGs. These two types of MGs are discussed below in detail.

1.4.3.1 Urban MGs

Urban MGs are established in urban areas close to the utility system and have the capability to operate in both IS mode and GC mode. The urban MGs follow all the rules, standards, synchronization techniques, and control strategies to maintain high power quality and ensure the stability of the UG system [32]. Universities, communities, hospitals, shopping malls, and offices are among the residential and commercial sectors where the urban MGs can be developed and implemented.

1.4.3.2 Remote MGs

More than 1 billion of the world's population living in underdeveloped and developing countries have no access to any reliable electricity. However, electricity is a main source for fulfilling basic human needs. In most cases, in these areas, the limited electricity that is accessible to these people is generated by using expensive fossil fuels, especially diesel fuel. Hence, MG can be considered the best possible solution to deliver electricity to these areas [33]. Remote MGs combine clean energy generation and ESS units, and in some scenarios, provide the facility of mobile payment platforms, providing a lifeline to the people, allowing the medical system to arrange reliable services, and allowing students to study during the night. Moreover, remote MGs are leveraging similar advances in communication and information technologies, PE technologies, and energy-generating technologies as compared to urban MGs. In remote MGs, the addition of RESs-based DG units with the existing diesel-based generating units shows great potential to lower the MG operating cost and diversify the generation in island communities that mainly rely on expensive oil for energy generation (due to being far away from the existing power system) [34].

1.4.4 Classification of MG Based on Scenario

Based on scenario, MGs can be classified into three categories, i.e., residential, industrial, and commercial MGs.

1.4.4.1 Residential MGs

A residential MG is composed of an advanced controller that combines the electricity demands of the consumer, regulates the DG, and coordinates with the distribution networks. This type of MG reduces the dependency of the consumer on electricity provided by the UG and also provides emergency power to the UG in case of a power outage. A controller in a residential MG turns a residence into a dynamic, flexible, and network resource that provides services to transmission and distribution network operators. This type of MG can be urban or rural in nature.

1.4.4.2 Industrial MGs

A key reason for implementing an industrial MG is to ensure reliable and secure power for the industry. In the production process of some industries, such as the chemical industry, chip production industry, food industry, and oil refinery, the power outage disturbs the production process due to lengthy start-up times of the machines, resulting in a considerable loss in revenue. Industrial MGs are now being installed in some industries because they offer numerous benefits compared to fossil fuel-based generating resources, such as integration of RESs and prosumer activity can be performed.

1.4.4.3 Commercial MGs

Commercial MGs are usually deployed to serve a single entity, such as data centres, hospitals and airports. This type of MG system is usually self-sufficient and can operate independently from the UG or operate in a GC mode at the time of need. Having the advantage of backup power, having the option to generate electricity from multiple resources, having day-a-head energy prices, and having the facility to connect to the UG can boost the self-independency and sufficiency of MG.

1.4.5 Classification of MG Based on Source

One of the most significant parts of the electrical system is the energy-generating resources. An MG is the combination of different energy-generating resources; therefore, to balance the demand-supply graph, numerous combinations of these resources can be used. From small to medium-scale MGs, DG units can be placed close to the utility premises or end-users to provide energy locally. Moreover, this DG-based technology can provide energy to remote and rural areas when the national transmission and distribution infrastructure is expensive to build [35]. Hence, based on the sources, the MGs can be classified into three categories, i.e., fossil fuel-based MGs, renewable-based MGs, and HMGs.

1.4.5.1 Fossil Fuel-Based MGs

In fossil fuel MGs, the generators are used to balance the demand-supply mismatch in remote areas. Before the advancement of RES-based generating technologies, energy resources like diesel generators and gas/steam turbines were used for power generation in remote MGs close to end-users. However, the power generated from these sources has a negative effect on both the economy and the environment. Moreover, the energy generated from fossil fuels is very expensive to transport and purchase, and, its transportation has its own carbon footprint. Hence, fossil fuel-based MGs are not the best solution to produce electricity in today's world, and researchers are exploring more sustainable and clean sources of energy due to these impacts [36].

1.4.5.2 RESs-Based MGs

The type of MGs that are powered by RESs-based DG units are known as RESs-based MGs. This type of MG usually consists of RESs and ESS units [37]. Compared to fossil fuels, the integration of RESs-based DGs in MG has increased significantly due to its sustainable and clean nature, low carbon emissions, and green governance attitude and policies. However, besides the numerous advantages, a main disadvantage of RES-based MGs is their vast variability and volatility in power generation due to their high dependency on meteorological factors. The intermittent and uncertain output of RES-based DGs increases the operational complexity of the MG [38]. Moreover, the time-varying energy demand also increases the operational complexity of an IS MG. Hence, to balance the demand-supply graph, an ESS is the most appealing and best solution [39]. An overview of RESs-based DG units, including their benefits, drawbacks, solutions, source of origin, and governance approach, is presented in Figure 1.13.

RESs-based MGs can be classified into five categories, namely, wind, solar, micro-hydro, biomass, and HMGs, that are discussed below.

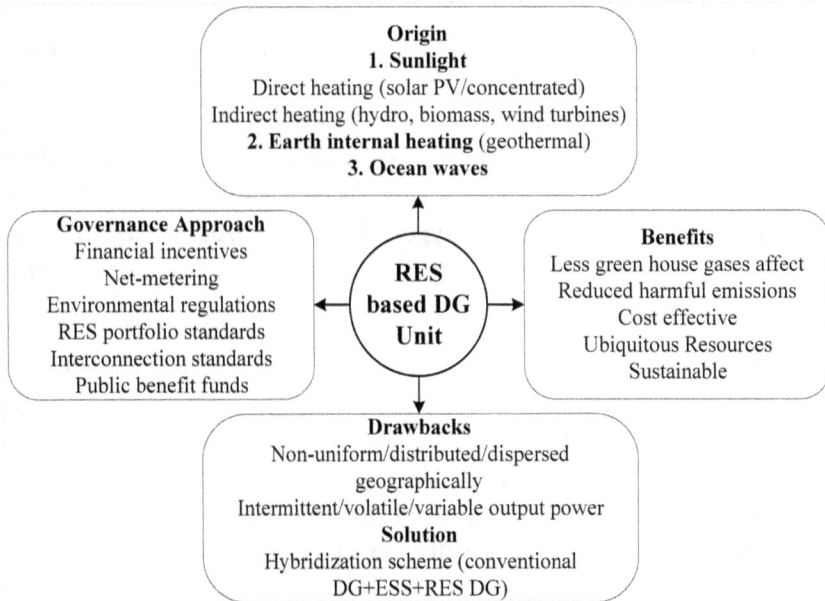

FIGURE 1.13   Overview of RESs-based DG units.

- **Wind Energy-Based MGs:** A wind energy-based MG can be defined as a distributed electrical system with a set of interlinked wind turbines and loads that operate as a single controllable entity within defined electrical boundaries. Due to the intermittent nature of wind, an ESS is usually employed in these types of MGs to store the extra generated energy for future use and provide an uninterruptible and smooth power supply. In Ref. [40], a wind energy-based MG is proposed and Monte Carlo framework-based simulations are performed to investigate the sensitivity of the system. In this work, it is concluded that the large ESS is also needed with the wind generation-based DG units of an MG due to the varying output power of the wind generation.

- **Solar Energy-Based MGs:** Solar-based MGs are an attractive solution to meet the ever-increasing electricity demand because they can be installed at any level and then scaled up at any time. Hence, they are considered the most durable and feasible electrification choice for rural/remote areas across the world. As solar-based MGs depend on solar irradiance for electricity generation; therefore, they should be installed in areas with an abundance of sunshine. However, due to cloud cover or partial shading conditions, the performance of the MG can be distorted; therefore, ESS must be installed in these MGs to ensure smooth performance [41]. Moreover, due to varying environmental conditions, the output power is fluctuating in nature; therefore, a suitable maximum power point technique and a robust controller should be used to extract the maximum power and ensure the system's stability. These types of MGs are usually used to power street lights, businesses, communities, hospitals, schools, irrigation pumps, etc.

- **Micro-Hydro-Based MGs:** These MGs are usually run-of-river projects that are installed on the river or stream. In a micro-hydro plant, the water is redirected from the stream or river into the turbine to generate electrical energy. In these power plants, the cost of generation is very low. However, the micro-hydro power plants are only limited to places that have sufficient water to run the turbine.

- **Biomass-Based MGs:** In this type of MG, an operation starts with the incomplete burning of any organic, decomposable matter derived from plants or animals in a gasifier to produce syngas. The syngas is then burned in an engine to power a generator [42]. Biomass-based MGs are gaining significant traction in some locations as the biomass gasifier is less expensive and the capital requirement is lower compared to solar PV systems. However, these MGs are only limited to locations that have enough biomass sources, as they require a large amount of feedstock, manpower, and storage procedures as compared to the other types of MGs. Moreover, bottle coil failure, battery discharge, and spark plug failure are all frequent problems that arise in these systems. Furthermore, wet husk also disturbs the energy generation in these MGs, and keeping the husk dry during monsoon season is also a challenging task [43].

- **Hybrid MGs:** In an HMG, different RESs such as biomass, PV, wind, and micro-hydro are combined together to generate electricity [44].

1.4.5.3 Hybrid MGs

In an HMG system, energy generation from fossil fuel, RESs, and ESS units are combined [44]. In the literature, different combinations of energy resources are combined together to develop an HMG. For example, the authors in Ref. [45] proposed an HMG in which RESs-based DG units are combined with the conventional gas engine to compensate for the fluctuating supply-demand. A combined approach of using both micro-turbines and micro-hydro in an MG is presented in Ref. [46]. The studies show that this type of MG architecture can be provided to rural areas with relatively weak natural resources and serve as a building block for the expansion of the existing system. The authors in Ref. [47] proposed a mix of different RESs-based DG units in a small MG, and their allocation is conducted to balance the supply-demand mismatch. From the state-of-the-art literature review, it can be concluded that conventional energy-based DG units, such as diesel generators, play a very important role in modern power systems due to the drawbacks related to RESs-based DGs. On the contrary, the advancement and low investment/cost of PV and wind technologies make them a very hot topic in the literature. However, compared to wind and PV very limited studies are conducted on other types of DGs such as fuel cells, geothermal, thermal power, concentrated solar power, and solid waste.

1.4.6 Classification of MG Based on Application

The MGs can be classified into numerous categories based on the application, such as campus, community, and military MGs.

1.4.6.1 Campus/Institutional MG

College campuses, universities, and corporations are all examples of this MG. These MGs are usually deployed on-site and are specialized in combined cooling, heating, and power applications.

1.4.6.2 Community MG

A community MG is supported by high penetration of RESs-based DGs and is frequently installed in the developed areas to support the communities in reaching the RE goals. This type of MG is limited to a specified electrically bound region and has the capability to operate in IS as well as in GC mode.

1.4.6.3 Military MG

The military MGs are small-scale electrical systems that are confined to the military base camps only and are mostly operated autonomously.

1.4.6.4 District Energy MG

These MGs provide both thermal energy and electricity for various facilities, such as heating and cooling.

1.4.6.5 Space MG

A reliable power system plays a significant role in the success of very expensive space missions; therefore, space MGs are designed to provide sustainable and reliable energy to meet the demands of satellites or spaceships [48].

1.4.6.6 Island MG

These are the small-scale MGs that work completely independently from the UG and generate their own power to meet energy demands.

1.4.6.7 Aerospace MG

In recent years, the concept of aerospace MG has gained great importance due to its increasing applications in aerospace technology. In numerous aerospace applications, such as electric aircrafts and airports, the pneumatic and mechanical power sources are gradually replaced by electrical sources [49].

1.4.6.8 Maritime MG

Maritime MGs are installed in ferries, ships, vessels, etc., and have the ability to work in IS mode at sea and in GC mode at port. The installation of these MGs is increasing rapidly as these MGs are affordable and the ships become more and more electrical [50].

1.4.7 Classification of MG Based on Size

Based on size, the MGs can be classified into three types, i.e., large, medium, and small-scale MGs [51].

1.4.7.1 Large-Scale MGs

The generation capacity of these MGs is greater than 100 MW and uses RESs/coal/oil or any combination of these electricity generating sources [52,53]. Large-scale MGs are capable of meeting the load demands of the industrial zone site.

1.4.7.2 Medium-Scale MGs

These MGs generate electricity of medium capacity by using RESs/coal/ oil or any combination of these sources. According to Refs. [52,53], the generating capacity of these MGs is greater than 10 MW and less than 100 MW. These MGs have the capability to meet the energy demands of industrial zones.

1.4.7.3 Small-Scale MGs

The electricity generating capacity of these MGs is low, and they use RESs for energy generation purposes. However, in some MGs, diesel generators are used along with the RESs or in place of RESs. The generating capacity of these MGs ranges up to 10 MW [35,36]. These MGs are capable of meeting the energy demand of remote areas, small residential communities, residential buildings, etc.

The prominent features such as application, generating capacity, and type of fuel of large, medium, and small-scale MGs are summarized in Table 1.2.

TABLE 1.2   Classification of MG Based on Size

Type	Generation Capacity (MW)	Application	Fuel
Large	>100	Industrial site	RESs/fuel/coal
Medium	10–100	Industrial zone	RESs/fuel/coal
Small	10	Remote area, island, residential building, small regional power grid	RESs

1.5 COMPONENTS OF MG

The major components that are involved in an MG are generation, ESS, energy management system (EMS), controller, loads, and PCC. These components will be discussed in this section.

1.5.1 Generation

The generation system of an MG consists of numerous dispatchable and non-dispatchable units. The dispatchable energy-generating units, such as diesel generators, biogas generators, and natural gas generators, are fully controlled units [54]. On the contrary, non-dispatchable generating units, including RESs such as wind and solar, are non-controllable due to their dependency on weather conditions [55].

1.5.2 Energy Storage System

Due to the intermittent and unpredictable power generation of RESs, the presence of ESS units is mandatory in the MG system. An ESS system in an MG not only smoothens the outputs but also performs numerous other functions such as frequency regulation, providing backup, ensuring high power quality, balancing the supply-demand graph, providing assistance in islanding, reducing operation, and optimization costs [56,57].

Based on the recent state-of-the-art literature, it is concluded that most of the ESS units are lithium-ion batteries, lead-acid batteries, and Hybrid Electric Vehicles (HEV)/Electric Vehicles (EV). The battery technologies used in HEV/EV can be utilized for managing and optimizing the operation of RESs-based DGs [58], demand response (DR), optimal parking lot [59], and vehicle-to-home [60], load management [61], frequency regulator, and harmonic compensator [62]. However, as compared to lithium-ion batteries, lithium-iron phosphate and zinc batteries offer numerous advantages. Zinc batteries have a lower cost due to the absence of flammable electrolytes; moreover, these are safer, have a higher energy density, and have a higher round-trip efficiency, thus providing more reliable and efficient energy storage. Furthermore, compared to lithium-ion batteries, lithium phosphate batteries also offer numerous advantageous features such as improved safety, a lower risk of thermal runaway, stable performance under high temperatures, and a longer cycle life. Based on these advantages, zinc and lithium phosphate batteries can be used as an alternative to lithium-ion batteries to overcome the drawbacks associated with it such as high cost, flammable electrolytes, and limited safety [63].

The advantages and disadvantages of the commonly used ESS units and generators in an MG are summarized in Table 1.3.

TABLE 1.3 Overview of MG Generation and ESS Options

Category	Options	Advantages	Disadvantages
Generation	Wind [64]	• Low energy-generating cost • Zero fuel cost • Low carbon footprint	• Noise and visual pollution • Reliance on wind • High capita and transportation cost
	Solar [65]	• Low carbon footprint • Zero fuel cost • Low maintenance cost • Diverse applications	• High capital cost • Requires ESS • Reliance on sun • High PE interface required
	Biogas [66]	• Reduces water pollution • Reduces soil pollution • Low cost of fuel source • Byproduct-fertilizer	• Fuel filtration • Fuel treatment • Require large reservoirs for continuous operation • Integration cost
	Fuel cells [67]	• Extremely quiet • Low emissions • CHP capable • Higher efficiency compared to micro-turbines	• Limited lifetime • Expensive hydrogen extraction • Expensive infrastructure of hydrogen is required
	Mini-hydro/ Micro-turbines [68]	• CHP capable • Dispatchable • Mechanical simplicity • Low emissions	• Greenhouse gas emission • Noise pollution
	Diesel Generator [69]	• Dispatchable • Fuel storage • High load acceptance • Fast transient response • Quick start-up	• High emissions • High fuel cost
	Gas Generator [70]	• Low emission • CHP capable • High fuel efficiency as compared to diesel generators	• Slow start-up • Slow transient response • High fuel price • Expensive fuel storage
Storage	Kinetic Energy Storage (Flywheels) [71]	• High efficiency • Fast response • High charge and discharge cycle	• High losses • Limited discharge time

(Continued)

TABLE 1.3 (*Continued*)   Overview of MG Generation and ESS Options

Category	Options	Advantages	Disadvantages
	Batteries (lithium ion, lead acid, nickel-cadmium, etc.) [72]	• History of development and research • Power availability is instant • Retrofit-able • Carbon saving neutrality	• Limited energy storage • Space constraint • Waste disposal • Battery life
	Hydrogen from hydrolysis [73]	• Clean	• Hydrogen storage is challenging • Low end-to-end efficiency
	Flow batteries "regenerative fuel cells" (bromide, vanadium redox zinc-bromine, etc.) [72]	• Decouple energy and power storage • Able to continuously support the maximum load and complete discharge with no damage risk	• Relatively early stage of deployment

1.5.3 Loads

For the reliable operation of a power system, a balance between energy demand and scheduled generation plays a very significant role. With the development of demand-side management programmes and the advent of metering infrastructure, the loads in an MG can also contribute to energy management. This concept is generally referred to as demand/load response. The main advantages of these programmes include enhanced transmission and distribution investment, large reserves for continuing generation, and improved efficacy and operational efficiency [74].

Hence, based on the concept of demand/load response management, the consumers who participate in this programme have their loads transformed into sinks or smart resources that are controllable. Another service provided by this concept is grid balancing, which is done either by load shaving or by load shifting in real time [75]. Besides these ancillary services provided by load management programmes, optimal load distribution, improved reliability, reduced emissions of greenhouse gases (GHG), and low cost. The loads in an MG can be divided into four categories: (a) important loads include critical (important to serve such as military and hospital) and non-critical (such as household) loads [76,77]; (b) consumption loads include industrial, commercial, and residential loads [78]; (c) responsive nature loads include curtailable and flexible loads (such as electric vehicle and ventilation system) [78]; and (d) non-responsive nature or un-controllable fixed loads include those loads in which there is no CN between the UG and the consumers [79].

1.5.4 Control in MG

MG offers several benefits, such as increasing the power system's reliability, reducing peak demand, reducing GHG emissions, having a prosumer-friendly architecture, integrating RESs, having a scalable nature, reducing energy losses, providing ancillary services, and improving power quality. Due to these advantages, an MG has gained significant interest in the past decade. However, besides these numerous advantages, there are some issues that need to be addressed, such as frequency deviation, system instability, voltage fluctuations, and supply-demand imbalances [80]. Due to these uncertainties, an MG is operated under highly stressed conditions; thus, a robust control strategy is required to guarantee an efficient and reliable operation [81].

The control of an MG is generally done in a hierarchical manner that consists of three levels, i.e., primary control level, secondary control (SC) level, and tertiary control level, and can be employed using three different approaches, i.e., centralized, decentralized, and distributed. Primary control is the basic level having the minimum decision time step and is concerned with the MG stability and power sharing within the network. The main objective of the SC is to eliminate the voltage and frequency deviations caused by the primary controller. Moreover, recently, more control objectives such as voltage unbalance, harmonic compensation, and reactive power sharing have been introduced in the additional SC loops [80]. The final level of the hierarchy is tertiary control, which is concerned with global economic optimization depending on energy prices and current markets [81].

1.5.5 Energy Management System in MG

An EMS in an MG revolves around an automated system that considers the MG operating modes, variability of energy resources, and operational cost and aims to optimally schedule the energy generation resources and ESS based on optimized control techniques. To achieve these goals, different researchers have developed different control approaches, such as the authors in Ref. [82] use a linear programming method to efficiently solve the cost function relative to the economical and technical operation of peak loads and DGs. Similarly, the authors in Ref. [83] use a genetic

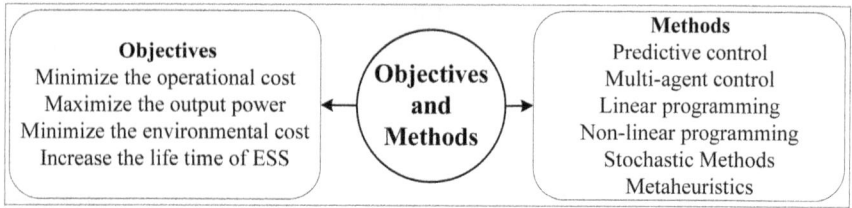

FIGURE 1.14   Objectives and EM methods for MG.

algorithm to optimally manage an HMG having a main objective is to minimize the operation cost by storing the extra energy in hydrogen form or in batteries.

A Java platform-based multi-agent system is designed for the outage condition of the host grid, in which the main objective is to minimize the operating cost while considering the price deviation of the host grid, variations of critical loads, and intermittency of DGs [84]. A control approach in which a stochastic scheme is combined with integer linear programming in order to minimize the operating cost and system losses of hybrid MG is presented in Ref. [85]. A two-layer model predictive controller-based EMS for hybrid MG is proposed in Ref. [86]. The main objective of this research is the inclusion of the degradation costs of supercapacitors and batteries for the precise evaluation of MG operating costs. Besides the above discussion, numerous other optimization techniques are present in the literature that are used for mono or multiple objectives based on optimization problems, such as operation, maintenance, cost of fuel, storage, and minimization of emissions. The main objective functions and the optimization methods used in MG are presented in Figure 1.14.

1.5.6 Point of Common Coupling

A PCC is a physical interlinking bridge between MG and UG. A PCC plays a very important role in the exchange of power between MG and UG. It comprises numerous devices and equipment (protective relays, circuit breaker, and synchronization equipment) that ensure a smooth connection of IS MG with the UG or vice versa and facilitate control, protection, and power exchange [87].

1.5.7 Power Conversion System

PE plays an important role in the field of distribution systems, RESs, MGs, telecommunication systems, traction fields, etc., due to its energy conversion capability. In all these areas, PE-based conversion systems boost the performance behaviour, efficacy, and development of the system. Moreover, in recent years, drastic improvements in PE technology have enhanced the dynamic performance and operating range while reducing the line harmonics in the systems. Besides these enhancements, in an MG, PE converters are also used to integrate different distributed energy sources like AC sources (wind, micro-turbines) and DC sources (PV, fuel cells, etc.) with the electrical system. For this purpose, different converters are used, such as AC-DC converters (rectifiers), DC-DC converters, and DC-AC converters (inverters).

1.5.7.1 Rectifiers

Rectifiers are used to convert AC voltage into DC voltage. These are widely used in MGs, such as in a DC MG system, where rectifiers are used to connect the AC generating source (such as wind) with the DC bus. The rectifiers are also used in a scenario where a DC load is connected to an AC bus in the MG system. In this scenario, an AC voltage is converted to a DC voltage to make it feasible for the required application.

1.5.7.2 DC-DC Converters

DC-DC step-up (boost) or step-down (buck) converters are used to integrate the DC generating source (such as PV) with the DC bus. The functionality of the DC-DC converter is not only limited to step-up or step-down of the voltage level but it also has to extract the maximum power from the PV system through the maximum power point technique and provide a regulated voltage at the output [88]. Similarly, to integrate the ESS (such as batteries) with the DC bus, bidirectional DC-DC converters are used that allow the direction of power flow in both directions.

1.5.7.3 Inverters

Inverters are used to convert the DC voltage into an AC voltage. In DC MG, the inverters are used between the DC bus and AC load to convert the voltage into AC to make it feasible for the AC load application. In AC MG, the inverters are used to integrate the DC-generated power source (such as PV, fuel cell) with the AC bus [89]. Moreover, due to the unpredictable and stochastic nature of RESs, especially PV, the inverter's functionality is not limited to voltage conversion but some intelligent and advanced ancillary functions such as power sharing, voltage regulation, frequency restoration, and harmonic minimization are also involved in its functionality [90].

1.6 EXAMPLE OF PV SYSTEM INTEGRATED WITH DC MG

Among the RESs, PV shows the fastest growth in the last consecutive years as a result the total installed capacity of PV reached from 305 GW in 2016 to 1185 GW in 2022 [11]. However, integration of PV with the MG is a challenging task due to their intermittent power generation and low generation voltage [91]. To cope with these issues, usually DC-DC converter is used that is equipped with the Maximum Power Point Tracking (MPPT) technique. In most literature, a conventional Perturb and Observe (P&O) algorithm and DC-DC converter are used to extract the maximum available power from the PV panel and step-up the low generated voltage to a high level. However, this is not a feasible solution as a conventional P&O technique is unable to accurately locate the maximum power point in case of varying environmental conditions; moreover, it results in steady-state oscillations and has slow dynamic response. Moreover, the usage of conventional DC-DC boost converters also poses numerous challenges such as (a) to attain high conversion, they have to be operated at a high duty cycle, and as a result, the efficiency of the system decreases due to parasitic resistance losses of diode, inductor, and capacitor; and (b) as a duty cycle increase, the voltage stress on the switch increases, which results in high conduction losses [92].

To cope with these challenges, an improved P&O technique is used in which the reference voltage acts as a function of temperature and irradiance that increases the accuracy of the algorithm over a wide spectrum of irradiance and temperature profiles [88]. Moreover, to step-up the conversion ratio, a new single-switch DC-DC converter was presented that has high structure modularity, simple design, and high efficiency [93]. A generalized schematic of this system is shown in Figure 1.15.

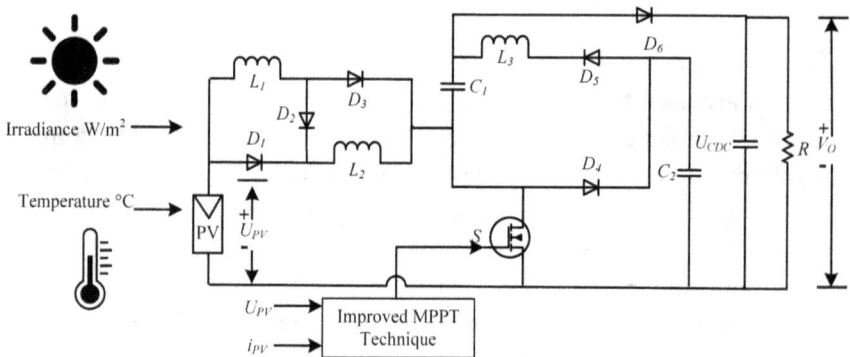

FIGURE 1.15   Schematic of PV system with DC-DC converter.

The performance of the system is validated under standard test conditions (irradiance $= 1000\,\text{W/m}^2$ and temperature $= 25°\text{C}$). An improved P&O technique efficiently tracks the MPP, and within 0.2, its comes to its steady state; moreover, the converter step-up the 17 V PV generated voltage to 108 V with the duty cycle of 0.73 (having a conversion ratio of 6.35) as shown in Figure 1.16a. The output power waveforms under standard test conditions are presented in Figure 1.16b.

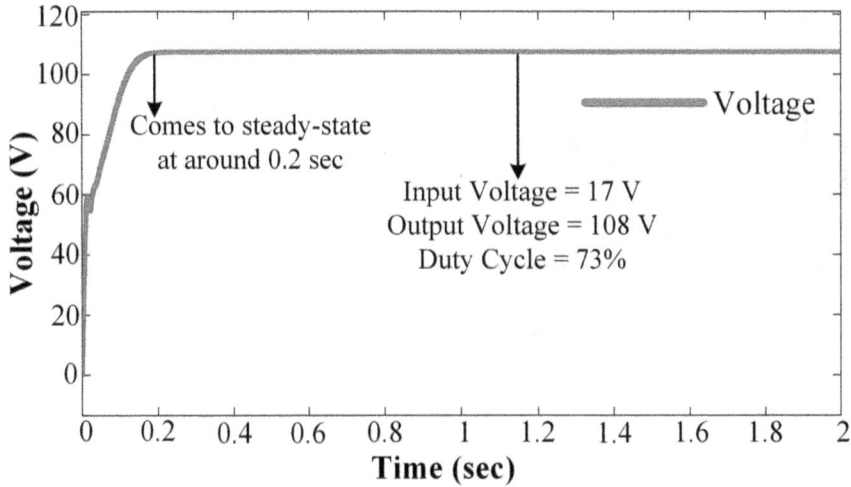

FIGURE 1.16a   Output voltage of a PV system under standard test conditions.

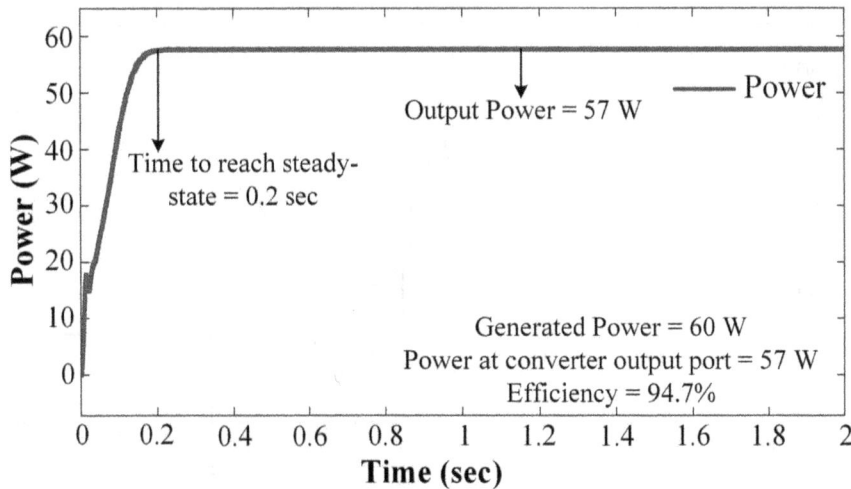

FIGURE 1.16b   Output power of a PV system under standard test conditions.

The performance of controller is tested in a scenario when both temperature and irradiance suddenly change at every 0.5 sec as shown in Figure 1.17a and b, respectively. The output power waveforms of both improved and conventional P&O techniques are presented in Figure 1.17c. From Figure 1.17c, it can be observed that the conventional P&O technique is sometimes unable to accurately locate the MPP and

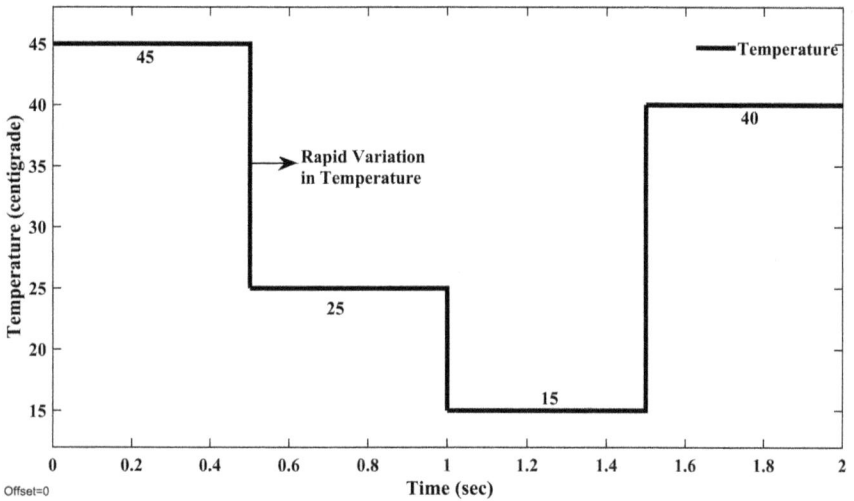

FIGURE 1.17a   Change in temperature.

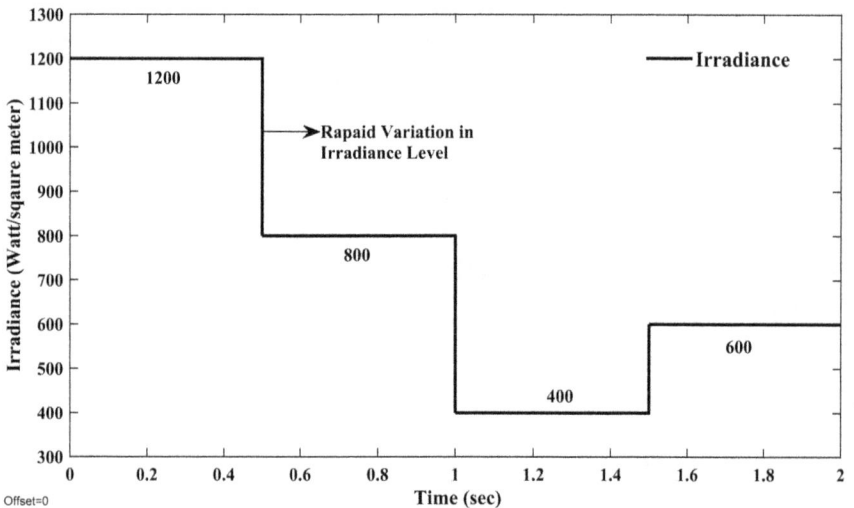

FIGURE 1.17b   Change in irradiance.

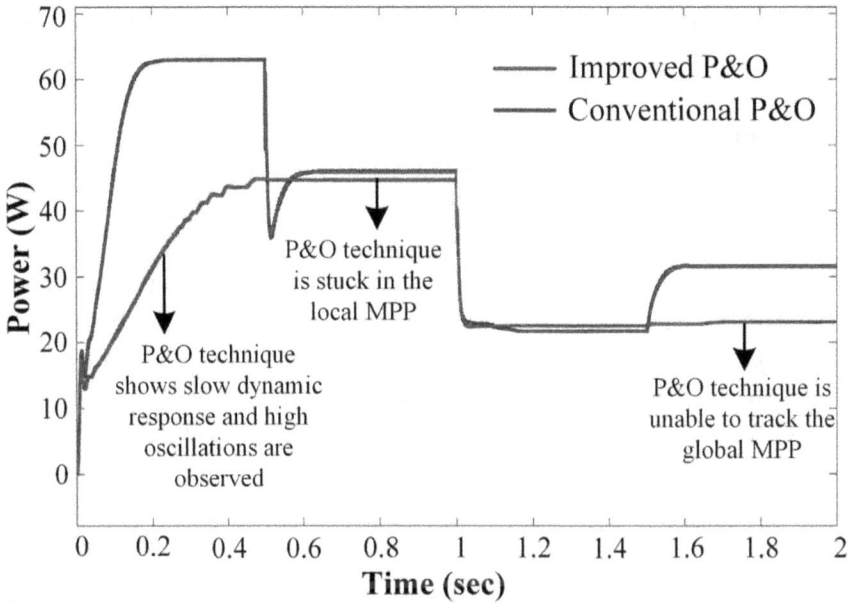

FIGURE 1.17c   Output power of PV system.

suffers from high steady-state oscillations; as a result, efficiency of the system is affected. Moreover, compared to the conventional P&O technique, an improved technique shows very effective performance and accurately locates the MPP with fast dynamic speed and low steady-state oscillations.

From this example, it can be concluded that the proper selection of power electronics converters and the maximum power extraction technique from the RESs plays a very important role in the efficient and reliable development of MG.

1.7 EXAMPLE OF CONNECTION OF DC GENERATING RESOURCES WITH AC MG

Referring to example 1.6, the PV system discussed above can be directly connected to the DC bus of the DC MG, while in case of AC MG, a DC-AC inverter is required to perform the inversion of DC voltage to an AC voltage. In literature, most of the authors used a conventional 2-level inverter for the inversion purpose. However, in an MG, the functionality of an inverter is not only limited to conversion but also includes control functionalities. Hence, a proper selection of an inverter topology plays a very significant role in the efficient and effective performance of an MG.

Let's consider a scenario in which 04 DC energy-generating units are required to connect to an AC MG. The use of an individual inverter for every DC energy-generating unit is not a feasible solution. Hence, to cope with this issue, a reduced switch asymmetric 31-level inverter that is capable to integrate DC generating units of different voltages with the UG or the PCC. A generalized schematic of 31-level inverter is shown in Figure 1.18 [90].

From Figure 1.18, it can be seen that different energy resources are connected to a single inverter through DC-DC converter. Moreover, an AC energy-generating source such as wind power system can also be connected to an inverter through a rectifier. Compared to a conventional 2-level inverter, this inverter is capable of integrating different sources with improved efficiency, reduced switching devices, increased reliability, and low cost. Besides the inverter topology, a selection of proper pulse

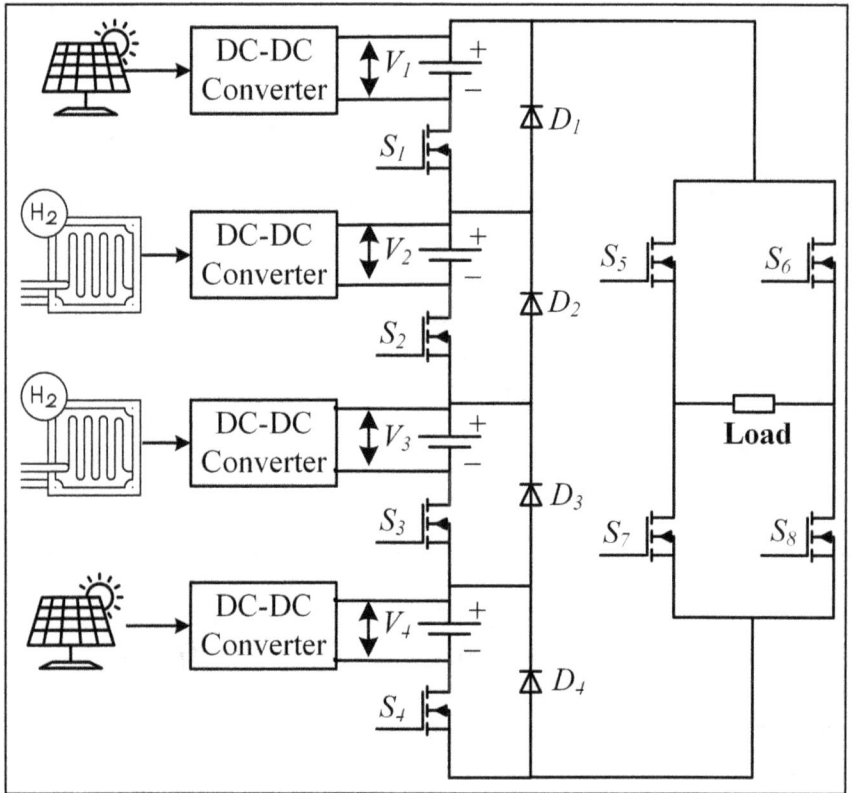

FIGURE 1.18 Schematic of inverter topology.

width modulation (PWM) scheme that drives the switches of the inverter also plays a very important role in the efficient performance of the MG system. Therefore, phase disposition PWM technique is used to drive the switches, as shown in Figure 1.19.

The performance of the inverter is tested for different DC inputs, and smooth output voltage and current waveforms are attained at the output, as can be seen in Figure 1.20a and b, respectively. Similarly, total harmonic distortion of 4.77% and 3.4% in output voltage and current waveforms (selected only 25 cycles) are observed, as shown in Figure 1.21a and b, respectively.

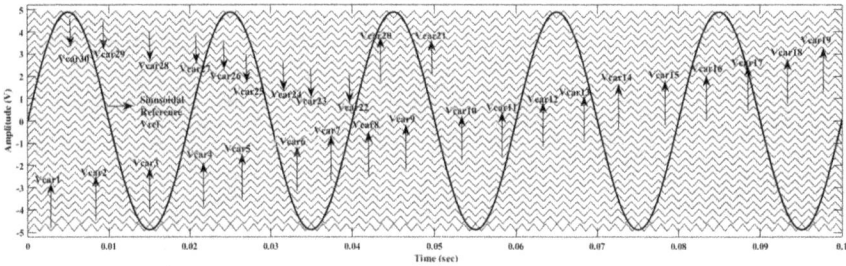

FIGURE 1.19 Phase disposition PWM technique.

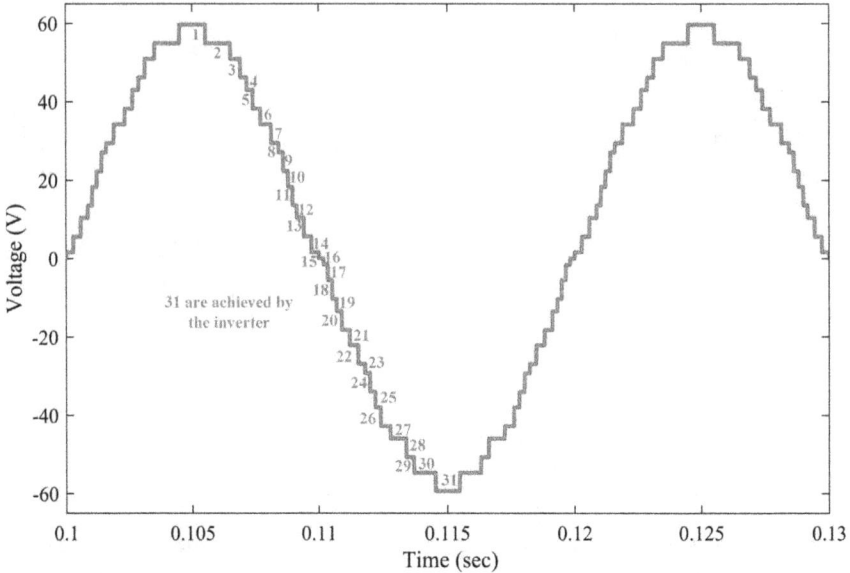

FIGURE 1.20a Output voltage waveform of inverter.

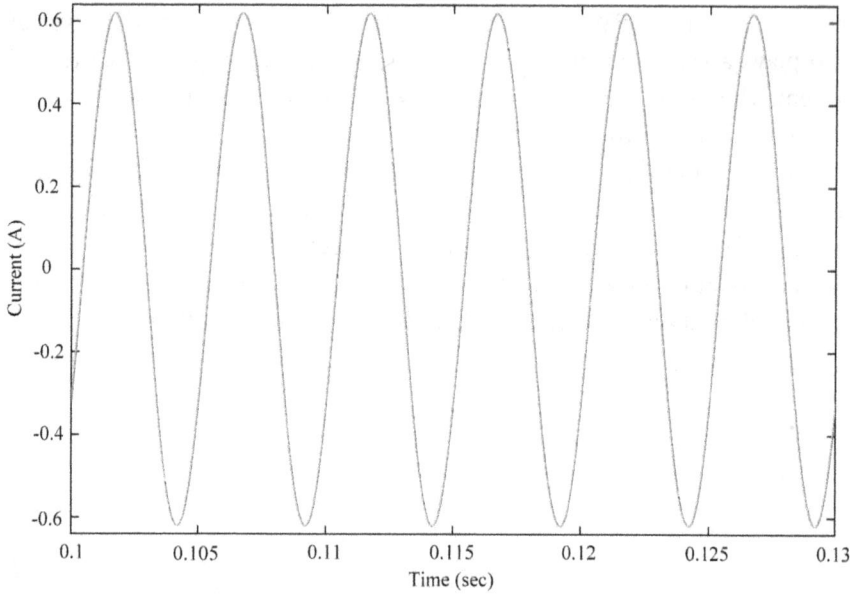

FIGURE 1.20b   Output current waveform of inverter.

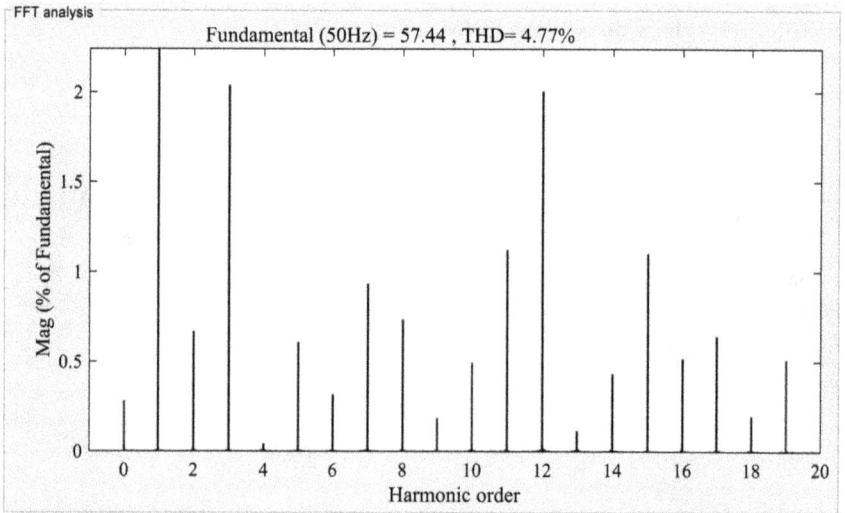

FIGURE 1.21a   THD results of output voltage waveform.

From the above discussion, it can be concluded that the accurate selection of inverter topology and PWN scheme is very important for the reliable and efficient performance of the system, as the whole operation of an

FFT analysis

Fundamental (50Hz) = 0.5732 , THD= 3.40%

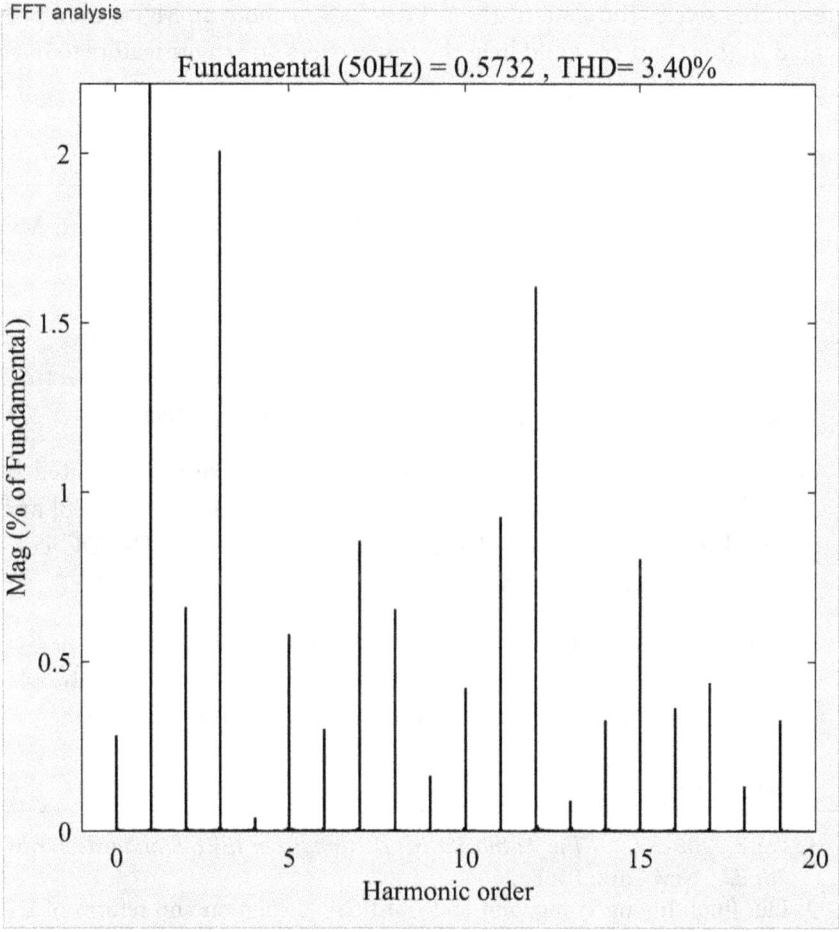

FIGURE 1.21b   THD results of output current waveform.

MG depends on the inverter. The advantages, limitations, and disadvantages of different control strategies that can be applied to inverter can be discussed in the upcoming chapters.

1.8 CONCLUSION

In this chapter, the fundamental concepts of an MG system, along with its different operating modes, are discussed. Furthermore, this chapter classifies an MG system into seven groups based on the end-user requirements, infrastructure, and application, and a detailed discussion of these groups is presented. Moreover, the main and fundamental components of an MG system, along with their different types and technologies, are discussed comprehensively. The state-of-the-art discussion about an MG system presented in this chapter would help the researchers and investigators to have a clear understanding of the basics of an MG system.

1.9 EXERCISES

1. What is the difference between traditional power system and MG system?

2. Summarize the different classifications of an MG.

3. Explore the different converters which are used to interconnect AC sub-grid and DC sub-grid in hybrid microgrid architecture.

4. Consider a multiport bidirectional DC-DC converter presented in Ref. [94] while using an MPPT technique discussed in Ref. [88] and check the performance of the MPPT technique and DC-DC converter through simulation results.

5. Consider a multi-level inverter discussed above as presented in Ref. [90] and try to use different PWM techniques such level shift, optimal switching angle, phase shift, etc., and compare the results.

REFERENCES

1. IEEE. "*100-2000 – The Authoritative Dictionary of IEEE Standards Terms, 7th Ed.*" New York: IEEE, 2000.
2. Liu, Jinqi, Jihong Wang, and Joel Cardinal. "Evolution and reform of UK electricity market." *Renewable and Sustainable Energy Reviews* 161 (2022): 112317.
3. Lakervi, Erkki, and Edward J. Holmes. *Electricity distribution network design*. No. 212. England: IET, 1995.
4. Alsuwian, Turki, Aiman Shahid Butt, and Arslan Ahmed Amin. "Smart grid cyber security enhancement: Challenges and solutions—A review." *Sustainability* 14, no. 21 (2022): 14226.
5. Cakir, Maide, Ilyas Cankaya, Ilhan Garip, and Ilhami Colak. "Advantages of Using Renewable Energy Sources in Smart Grids." In *2022 10th International Conference on Smart Grid (ICSmartGrid)*, pp. 436–439. IEEE, Istanbul, Turkey, 2022.
6. Ali, Sahibzada Muhammad, Muhammad Jawad, Bilal Khan, Chaudhry Arshad Mehmood, Nadia Zeb, Ayesha Tanoli, Umar Farid, Jacob Glower, and Samee U. Khan. "Wide area smart grid architectural model and control: A survey." *Renewable and Sustainable Energy Reviews* 64 (2016): 311–328.

7. Barker, Philip P., and Robert W. De Mello. "Determining the impact of distributed generation on power systems. I. Radial distribution systems." In *2000 Power Engineering Society Summer Meeting (Cat. No. 00CH37134)*, vol. 3, pp. 1645–1656. IEEE, Seattle, WA, USA, 2000.
8. Li, Wensheng, Rong Liang, Fengzhang Luo, Shaoting Feng, Bo Yang, Zhao Liu, Zhao Li, Wen Zhang, Kuangyu Kong, and Siyuan Lu. "Response potential assessment of user-side flexible resources of regional power distribution networks based on sequential simulation of optimal operation." *Frontiers in Energy Research* 10 (2022): 1096046.
9. Judge, Malik Ali, Asif Khan, Awais Manzoor, and Hasan Ali Khattak. "Overview of smart grid implementation: Frameworks, impact, performance and challenges." *Journal of Energy Storage* 49 (2022): 104056.
10. Vega-Garita, Victor, Anindio Prabu Harsarapama, Laura Ramirez-Elizondo, and Pavol Bauer. "Physical integration of PV-battery system: Advantages, challenges, and thermal model." In *2016 IEEE International Energy Conference (ENERGYCON)*, pp. 1–6. IEEE, Leuven, Belgium, 2016.
11. Gibb, Duncan, Nathalie Ledanois, Lea Ranalder, and Hend Yaqoob. "Renewables 2022 Global Status Report." REN21: Paris, France, 2022.
12. Zeb, Kamran, Saif Ul Islam, Waqar Ud Din, Imran Khan, Muhammad Ishfaq, Tiago Davi Curi Busarello, Iftikhar Ahmad, and Hee Je Kim. "Design of fuzzy-PI and fuzzy-sliding mode controllers for single-phase two-stages grid-connected transformerless photovoltaic inverter." *Electronics* 8, no. 5 (2019): 520.
13. Marzal, Silvia, Robert Salas, Raúl González-Medina, Gabriel Garcerá, and Emilio Figueres. "Current challenges and future trends in the field of communication architectures for microgrids." *Renewable and Sustainable Energy Reviews* 82 (2018): 3610–3622.
14. Farhangi, Hassan, ed. *Smart Microgrids: Lessons from Campus Microgrid Design and Implementation.* Boca Raton, FL: CRC Press, 2016.
15. Tabatabaei, Naser Mahdavi, Ersan Kabalci, and Nicu Bizon, eds. *Microgrid Architectures, Control and Protection Methods.* Cham, Switzerland: Springer, 2019.
16. Pogaku, Nagaraju, Milan Prodanovic, and Timothy C. Green. "Modeling, analysis and testing of autonomous operation of an inverter-based microgrid." *IEEE Transactions on Power Electronics* 22, no. 2 (2007): 613–625.
17. Khan, Muhammad Yasir Ali, Haoming Liu, Ren Zhang, Qi Guo, Haiqing Cai, and Libin Huang. "A unified distributed hierarchal control of a microgrid operating in islanded and grid connected modes." *IET Renewable Power Generation* 17, no. 10 (2023): 2489–2511.
18. Ray, Papia, and Monalisa Biswal. *Microgrid: Operation, Control, Monitoring and Protection.* Singapore: Springer, 2020.
19. Saeed, Laraib, Muhammad Yasir Ali Khan, Hamid Karim, and Emad Alhani. "A bidirectional DC-DC bipolar converter for power transmission network." In *2021 International Conference on Computing, Electronic and Electrical Engineering (ICE Cube)*, pp. 1–6. IEEE, Quetta, Pakistan, 2021.

20. Shah, Krushal, Penghao Chen, Adam Schwab, Krishna Shenai, Sebastien Gouin-Davis, and Liang Xi Downey. "Smart efficient solar DC micro-grid." In *2012 IEEE Energytech*, pp. 1–5. IEEE, Cleveland, OH, USA, 2012.
21. Khan, Muhammad Yasir Ali, Haoming Liu, Jie Shang, and Jian Wang. "Distributed hierarchal control strategy for multi-bus AC microgrid to achieve seamless synchronization." *Electric Power Systems Research* 214 (2023): 108910.
22. Gomez-Redondo, Marcos, Marco Rivera, Javier Muñoz, and Patrick Wheeler. "A systematic literature review on AC microgrids." *Designs* 8, no. 4 (2024): 77.
23. Dagar, Annu, Pankaj Gupta, and Vandana Niranjan. "Microgrid protection: A comprehensive review." *Renewable and Sustainable Energy Reviews* 149 (2021): 111401.
24. Hatziargyriou, Nikos, ed. *Microgrids: Architectures and Control*. Hoboken, NJ: John Wiley & Sons, 2014.
25. Shezan, S. K. Arefin, Innocent Kamwa, Md Fatin Ishraque, S. M. Muyeen, Kazi Nazmul Hasan, Rahman Saidur, Syed Muhammad Rizvi, Md Shafiullah, and Fahad A. Al-Sulaiman. "Evaluation of different optimization techniques and control strategies of hybrid microgrid: A review." *Energies* 16, no. 4 (2023): 1792.
26. Unamuno, Eneko, and Jon Andoni Barrena. "Hybrid ac/dc microgrids—Part I: Review and classification of topologies." *Renewable and Sustainable Energy Reviews* 52 (2015): 1251–1259.
27. Chen, Bo, Jianhui Wang, Xiaonan Lu, Chen Chen, and Shijia Zhao. "Networked microgrids for grid resilience, robustness, and efficiency: A review." *IEEE Transactions on Smart Grid* 12, no. 1 (2020): 18–32.
28. Ali, Naghmash, Xinwei Shen, Hammad Armghan, and Yunfei Du. "Hierarchical control combined with higher order sliding mode control for integrating wind/tidal/battery/hydrogen powered DC offshore microgrid." *Journal of Energy Storage* 82 (2024): 110521.
29. Sabri, Yassine, Najib El Kamoun, and Fatima Lakrami. "A survey: Centralized, decentralized, and distributed control scheme in smart grid systems." In *2019 7th Mediterranean Congress of Telecommunications (CMT)*, pp. 1–11. IEEE, Fez, Morocco, 2019.
30. Elbez, Ghada, Hubert B. Keller, and Veit Hagenmeyer. "A new classification of attacks against the cyber-physical security of smart grids." In *Proceedings of the 13th International Conference on Availability, Reliability and Security*, pp. 1–6. IEEE, Hamburg Germany, 2018.
31. Espina, Enrique, Jacqueline Llanos, Claudio Burgos-Mellado, Roberto Cardenas-Dobson, Manuel Martinez-Gomez, and Doris Saez. "Distributed control strategies for microgrids: An overview." *IEEE Access* 8 (2020): 193412–193448.
32. Chandak, Sheetal, Pritam Bhowmik, and Pravat Kumar Rout. "Load shedding strategy coordinated with storage device and D-STATCOM to enhance the microgrid stability." *Protection and Control of Modern Power Systems* 4, no. 3 (2019): 1–19.

33. Williams, Nathaniel J., Paulina Jaramillo, Jay Taneja, and Taha Selim Ustun. "Enabling private sector investment in microgrid-based rural electrification in developing countries: A review." *Renewable and Sustainable Energy Reviews* 52 (2015): 1268–1281.

34. Kuang, Yonghong, Yongjun Zhang, Bin Zhou, Canbing Li, Yijia Cao, Lijuan Li, and Long Zeng. "A review of renewable energy utilization in islands." *Renewable and Sustainable Energy Reviews* 59 (2016): 504–513.

35. Lee, Jun Yin, Renuga Verayiah, Kam Hoe Ong, Agileswari K. Ramasamy, and Marayati Binti Marsadek. "Distributed generation: A review on current energy status, grid-interconnected PQ issues, and implementation constraints of DG in Malaysia." *Energies* 13, no. 24 (2020): 6479.

36. Schnitzer, Daniel, Deepa Shinde Lounsbury, Juan Pablo Carvallo, Ranjit Deshmukh, Jay Apt, and Daniel M. Kammen. *"Microgrids for Rural Electrification."* New York: United Nations Foundation, 2014.

37. He, Li, Shiyue Zhang, Yizhong Chen, Lixia Ren, and Jing Li. "Techno-economic potential of a renewable energy-based microgrid system for a sustainable large-scale residential community in Beijing, China." *Renewable and Sustainable Energy Reviews* 93 (2018): 631–641.

38. Gao, David Wenzhong. *Energy Storage for Sustainable Microgrid.* New York: Academic Press, 2015.

39. He, Jinghan, Xiaoyu Wu, Xiangyu Wu, Yin Xu, and Josep M. Guerrero. "Small-signal stability analysis and optimal parameters design of microgrid clusters." *IEEE Access* 7 (2019): 36896–36909.

40. Giacomoni, Anthony M., Steven Y. Goldsmith, S. Massoud Amin, and Bruce F. Wollenberg. "Analysis, modeling, and simulation of autonomous microgrids with a high penetration of renewables." In *2012 IEEE Power and Energy Society General Meeting*, pp. 1–6. IEEE, San Diego, CA, USA, 2012.

41. Numminen, Sini, and Peter D. Lund. "Evaluation of the reliability of solar micro-grids in emerging markets–Issues and solutions." *Energy for Sustainable Development* 48 (2019): 34–42.

42. Liew, Peng Yen, Petar Sabav Varbanov, Aoife Foley, and Jiří Jaromír Klemeš. "Smart energy management and recovery towards sustainable energy system optimisation with bio-based renewable energy." *Renewable and Sustainable Energy Reviews* 135 (2021): 110385.

43. Kumar, Peddapelli Satish, Ruwan P. Sarath Chandrasena, Vasa Ramu, G. N. Srinivas, and K. Victor Sam Moses Babu. "Energy management system for small scale hybrid wind solar battery based microgrid." *IEEE Access* 8 (2020): 8336–8345.

44. Wang, Chengshan, Yixin Liu, Xialin Li, Li Guo, Lei Qiao, and Hai Lu. "Energy management system for stand-alone diesel-wind-biomass microgrid with energy storage system." *Energy* 97 (2016): 90–104.

45. Asano, Hiroshi, and Shigeru Bando. "Load fluctuation analysis of commercial and residential customers for operation planning of a hybrid photovoltaic and cogeneration system." In *2006 IEEE Power Engineering Society General Meeting*, pp. 6. IEEE, Montreal, QC, Canada, 2006.

46. Litifu, Zulati, Noel Estoperez, Mostafa Al Mamun, Ken Nagasaka, Yasuyuki Nemoto, and Izumi Ushiyama. "Planning of micro-grid power supply based on the weak wind and hydro power generation." In *2006 IEEE Power Engineering Society General Meeting*, pp. 8. IEEE, Montreal, QC, Canada, 2006.

47. Quiggin, Daniel, Sarah Cornell, Michael Tierney, and Richard Buswell. "A simulation and optimisation study: Towards a decentralised microgrid, using real world fluctuation data." *Energy* 41, no. 1 (2012): 549–559.

48. Lashab, Abderezak, Mohammad Yaqoob, Yacine Terriche, Juan C. Vasquez, and Josep M. Guerrero. "Space microgrids: New concepts on electric power systems for satellites." *IEEE Electrification Magazine* 8, no. 4 (2020): 8–19.

49. Tarisciotti, Luca, Alessandro Costabeber, Linglin Chen, Adam Walker, and Michael Galea. "Current-fed isolated DC/DC converter for future aerospace microgrids." *IEEE Transactions on Industry Applications* 55, no. 3 (2018): 2823–2832.

50. Jin, Zheming, Giorgio Sulligoi, Rob Cuzner, Lexuan Meng, Juan C. Vasquez, and Josep M. Guerrero. "Next-generation shipboard dc power system: Introduction smart grid and dc microgrid technologies into maritime electrical netowrks." *IEEE Electrification Magazine* 4, no. 2 (2016): 45–57.

51. Shahgholian, Ghazanfar. "A brief review on microgrids: Operation, applications, modeling, and control." *International Transactions on Electrical Energy Systems* 31, no. 6 (2021): e12885.

52. Hu, Jingwei, Tieyan Zhang, Shipeng Du, and Yan Zhao. "An overview on analysis and control of micro-grid system." *International Journal of Control and Automation* 8, no. 6 (2015): 65–76.

53. Singh, Niharika, Irraivan Elamvazuthi, Perumal Nallagownden, Gobbi Ramasamy, and Ajay Jangra. "Routing based multi-agent system for network reliability in the smart microgrid." *Sensors* 20, no. 10 (2020): 2992.

54. Mariam, Lubna, Malabika Basu, and Michael F. Conlon. "A review of existing microgrid architectures." *Journal of Engineering* 2013, no. 1 (2013): 937614.

55. Baker, Kyri, Gabriela Hug, and Xin Li. "Optimal integration of intermittent energy sources using distributed multi-step optimization." In *2012 IEEE Power and Energy Society General Meeting*, pp. 1–8. IEEE, San Diego, CA, USA, 2012.

56. Choudhury, Subhashree. "Review of energy storage system technologies integration to microgrid: Types, control strategies, issues, and future prospects." *Journal of Energy Storage* 48 (2022): 103966.

57. Uddin, Moslem, Mohd Fakhizan Romlie, Mohd Faris Abdullah, Syahirah Abd Halim, and Tan Chia Kwang. "A review on peak load shaving strategies." *Renewable and Sustainable Energy Reviews* 82 (2018): 3323–3332.

58. Khan, Muhammad Waseem, and Jie Wang. "Multi-agents based optimal energy scheduling technique for electric vehicles aggregator in microgrids." *International Journal of Electrical Power & Energy Systems* 134 (2022): 107346.

59. Mohan, Vivek, Jai Govind Singh, and Weerakorn Ongsakul. "Sortino ratio based portfolio optimization considering EVs and renewable energy in microgrid power market." *IEEE Transactions on Sustainable Energy* 8, no. 1 (2016): 219–229.

60. Rahman, Md Shihanur, and Aman Maung Than Oo. "Distributed multi-agent based coordinated power management and control strategy for microgrids with distributed energy resources." *Energy conversion and management* 139 (2017): 20–32.

61. Mbungu, Nsilulu Tresor, Tshimbalanga Madiba, Ramesh C. Bansal, Maamar Bettayeb, Raj M. Naidoo, Mukwanga Willy Siti, and Temitope Adefarati. "Economic optimal load management control of microgrid system using energy storage system." *Journal of Energy Storage* 46 (2022): 103843.

62. Siahroodi, Hossein Jafari, Hamed Mojallali, and Seyed Saeid Mohtavipour. "A novel multi-objective framework for harmonic power market including plug-in electric vehicles as harmonic compensators using a new hybrid gray wolf-whale-differential evolution optimization." *Journal of Energy Storage* 52 (2022): 105011.

63. Gao, Yan, Yingling Cai, and Chenglin Liu. "Annual operating characteristics analysis of photovoltaic-energy storage microgrid based on retired lithium iron phosphate batteries." *Journal of Energy Storage* 45 (2022): 103769.

64. Zhao, Zhuoli, Xi Luo, Jindian Xie, Shaoqing Gong, Juntao Guo, Qiang Ni, Chun Sing Lai, Ping Yang, Loi Lei Lai, and Josep M. Guerrero. "Decentralized grid-forming control strategy and dynamic characteristics analysis of high-penetration wind power microgrids." *IEEE Transactions on Sustainable Energy* 13, no. 4 (2022): 2211–2225.

65. Majji, Ravi Kumar, Jyoti Prakash Mishra, and Ashish A. Dongre. "Model predictive control of solar photovoltaic-based microgrid with composite energy storage." *International Journal of Circuit Theory and Applications* 50, no. 7 (2022): 2490–2509.

66. Planas, Estefanía, Jon Andreu, José Ignacio Gárate, Iñigo Martínez De Alegría, and Edorta Ibarra. "AC and DC technology in microgrids: A review." *Renewable and Sustainable Energy Reviews* 43 (2015): 726–749.

67. Akinyele, Daniel, Elijah Olabode, and Abraham Amole. "Review of fuel cell technologies and applications for sustainable microgrid systems." *Inventions* 5, no. 3 (2020): 42.

68. Wang, Gan, Zhifeng Chen, Zhidong Wang, Zhiheng Xu, Zifan Zhang, Yifeng Liu, and Xiaodie Zhang. "Research and implementation of frequency control strategy of islanded microgrids rich in grid-connected small hydropower." *Energy Reports* 9 (2023): 5053–5063.

69. Hirsch, Adam, Yael Parag, and Josep Guerrero. "Microgrids: A review of technologies, key drivers, and outstanding issues." *Renewable and Sustainable Energy Reviews* 90 (2018): 402–411.

70. Renjit, Ajit Anbiah, Mahesh Sitaram Illindala, Robert H. Lasseter, Micah J. Erickson, and David Klapp. "Modeling and control of a natural gas generator set in the CERTS microgrid." In *2013 IEEE Energy Conversion Congress and Exposition*, pp. 1640–1646. IEEE, Denver, CO, USA, 2013.

71. Kikusato, Hiroshi, Taha Selim Ustun, Masaichi Suzuki, Shuichi Sugahara, Jun Hashimoto, Kenji Otani, Nobuyoshi Ikeda, Iichiro Komuro, Hideaki Yokoi, and Kunihiro Takahashi. "Flywheel energy storage system based microgrid controller design and PHIL testing." *Energy Reports* 8 (2022): 470–475.

72. Abu-Sharkh, Suleiman, R. J. Arnold, Jonathan Kohler, R. Li, Tomas Markvart, J. Neil Ross, Koen Steemers, Peter Wilson, and Runming Yao. "Can microgrids make a major contribution to UK energy supply?." *Renewable and Sustainable Energy Reviews* 10, no. 2 (2006): 78–127.

73. Díaz-González, Francisco, Andreas Sumper, Oriol Gomis-Bellmunt, and Roberto Villafáfila-Robles. "A review of energy storage technologies for wind power applications." *Renewable and Sustainable Energy Reviews* 16, no. 4 (2012): 2154–2171.

74. Samad, Tariq, Edward Koch, and Petr Stluka. "Automated demand response for smart buildings and microgrids: The state of the practice and research challenges." *Proceedings of the IEEE* 104, no. 4 (2016): 726–744.

75. Imani, Mahmood Hosseini, M. Jabbari Ghadi, Sahand Ghavidel, and Li Li. "Demand response modeling in microgrid operation: a review and application for incentive-based and time-based programs." *Renewable and Sustainable Energy Reviews* 94 (2018): 486–499.

76. Vergara, Pedro P., Juan Camilo López, Luiz CP da Silva, and Marcos J. Rider. "Security-constrained optimal energy management system for three-phase residential microgrids." *Electric Power Systems Research* 146 (2017): 371–382.

77. Du, Yigao, Jing Wu, Shaoyuan Li, Chengnian Long, and Ioannis Ch Paschalidis. "Distributed MPC for coordinated energy efficiency utilization in microgrid systems." *IEEE Transactions on Smart Grid* 10, no. 2 (2017): 1781–1790.

78. Khalili, Tohid, Sayyad Nojavan, and Kazem Zare. "Optimal performance of microgrid in the presence of demand response exchange: A stochastic multi-objective model." *Computers & Electrical Engineering* 74 (2019): 429–450.

79. Carpinelli, Guido, Fabio Mottola, Daniela Proto, and Pietro Varilone. "Minimizing unbalances in low-voltage microgrids: Optimal scheduling of distributed resources." *Applied Energy* 191 (2017): 170–182.

80. Ali, Hossam, Gaber Magdy, and Dianguo Xu. "A new optimal robust controller for frequency stability of interconnected hybrid microgrids considering non-inertia sources and uncertainties." *International Journal of Electrical Power & Energy Systems* 128 (2021): 106651.

81. Som, Shreyasi, Souradip De, Saikat Chakrabarti, Soumya Ranjan Sahoo, and Arindam Ghosh. "A robust controller for battery energy storage system of an islanded AC microgrid." *IEEE Transactions on Industrial Informatics* 18, no. 1 (2021): 207–218.

82. Ahmad, Jameel, Muhammad Imran, Abdullah Khalid, Waseem Iqbal, Syed Rehan Ashraf, Muhammad Adnan, Syed Farooq Ali, and Khawar Siddique Khokhar. "Techno economic analysis of a wind-photovoltaic-biomass hybrid renewable energy system for rural electrification: A case study of Kallar Kahar." *Energy* 148 (2018): 208–234.

83. Dufo-Lopez, Rodolfo, José L. Bernal-Agustín, and Javier Contreras. "Optimization of control strategies for stand-alone renewable energy systems with hydrogen storage." *Renewable Energy* 32, no. 7 (2007): 1102–1126.

84. Raju, Leo, Antony Amalraj Morais, Ramyaa Rathnakumar, Soundaryaa Ponnivalavan, and L. D. Thavam. "Micro-grid grid outage management using multi-agent systems." In *2017 Second International Conference on Recent Trends and Challenges in Computational Models (ICRTCCM)*, pp. 363–368. IEEE, Tindivanam, India, 2017.

85. Reddy, S. Surender. "Optimization of renewable energy resources in hybrid energy systems." *J. Green Eng* 7, no. 1 (2017): 43–60.

86. Ju, Chengquan, Peng Wang, Lalit Goel, and Yan Xu. "A two-layer energy management system for microgrids with hybrid energy storage considering degradation costs." *IEEE Transactions on Smart Grid* 9, no. 6 (2017): 6047–6057.

87. Lidula, Nilakshi Widanagama Arachchige, and Athula Dayanath Rajapakse. "Microgrids research: A review of experimental microgrids and test systems." *Renewable and Sustainable Energy Reviews* 15, no. 1 (2011): 186–202.

88. Zamani, Mehdi, Amir Aghaei, Seyed Majid Hashemzadeh, and Mehran Sabahi. "Improved P&O Algorithm for Maximum Power Point Tracking at the Photovoltaic Array Using an Interleaved Boost Converter." In *2020 28th Iranian Conference on Electrical Engineering (ICEE)*, pp. 1–5. IEEE, Tabriz, Iran, 2020.

89. Liu, Haoming, Muhammad Yasir Ali Khan, and Xiaoling Yuan. "Hybrid maximum power extraction methods for photovoltaic systems: A comprehensive review." *Energies* 16, no. 15 (2023): 5665.

90. Manoranjan, A., and C. Christober Asir Rajan. "Design and analysis of 31-level asymmetric cascaded H-bridge multilevel inverter with reduced number of switches." *Bull. Sci. Res* 14, no. 28 (2020): 14.

91. Khan, Muhammad Yasir Ali, Muhammad Azhar, Laraib Saeed, Sajjad Ali Khan, and Jahangeer Soomro. "A High Gain Multiport Non-Isolated DC-DC Converter for PV Applications." In *2019 2nd International Conference on Computing, Mathematics and Engineering Technologies (iCoMET)*, pp. 1–6. IEEE, Sukkur, Pakistan, 2019.

92. Khan, Muhammad Yasir Ali, Haoming Liu, and Naveed Ur Rehman. "Design of a multiport bidirectional DC-DC converter for low power PV applications." In *2021 International Conference on Emerging Power Technologies (ICEPT)*, pp. 1–6. IEEE, Topi, Pakistan, 2021.

93. Khan, Muhammad Yasir Ali, Haoming Liu, Salman Habib, Danish Khan, and Xiaoling Yuan. "Design and performance evaluation of a step-up DC–DC converter with dual loop controllers for two stages grid connected PV inverter." *Sustainability* 14, no. 2 (2022): 811.

94. Azhar, Muhammad, Muhammad Yasir Ali Khan, Laraib Saeed, Jawad Saleem, and Abdul Majid. "Design and analysis of bidirectional sepic-based boost multi-port converter." In *2019 International Conference on Engineering and Emerging Technologies (ICEET)*, pp. 1–6. IEEE, Lahore, Pakistan, 2019.

CHAPTER 2

Benefits, Challenges, and Technical Aspects of Microgrid

2.1 BENEFITS OF MG

As compared to traditional power grids, MGs offer numerous benefits, as presented in Figure 2.1 and are discussed below in detail.

2.1.1 Environmental Benefits

As compared to conventional power generation, energy generation in MGs is usually based on RESs that have negligible environmental impacts. The energy generation from RESs helps in pollution reduction, reduction in harmful gases, reduction in GHG emissions, and improvement in the quality of the air. The emission of GHG and carbon has a significant impact on the increase in global temperature; therefore, the deployment of RESs-based MGs will contribute a lot to reduce the global temperature [1]. However, besides these numerous environmental benefits, there are some negative environmental impacts, such as in the case of biomass generation. A biomass energy generating unit requires a large area to use the feedstock production can, which in turn causes deforestation and other negative environmental impacts. Therefore, careful consideration should be given while deploying an MG to ensure that only positive environmental impacts are enhanced [2].

DOI: 10.1201/9781003594284-2

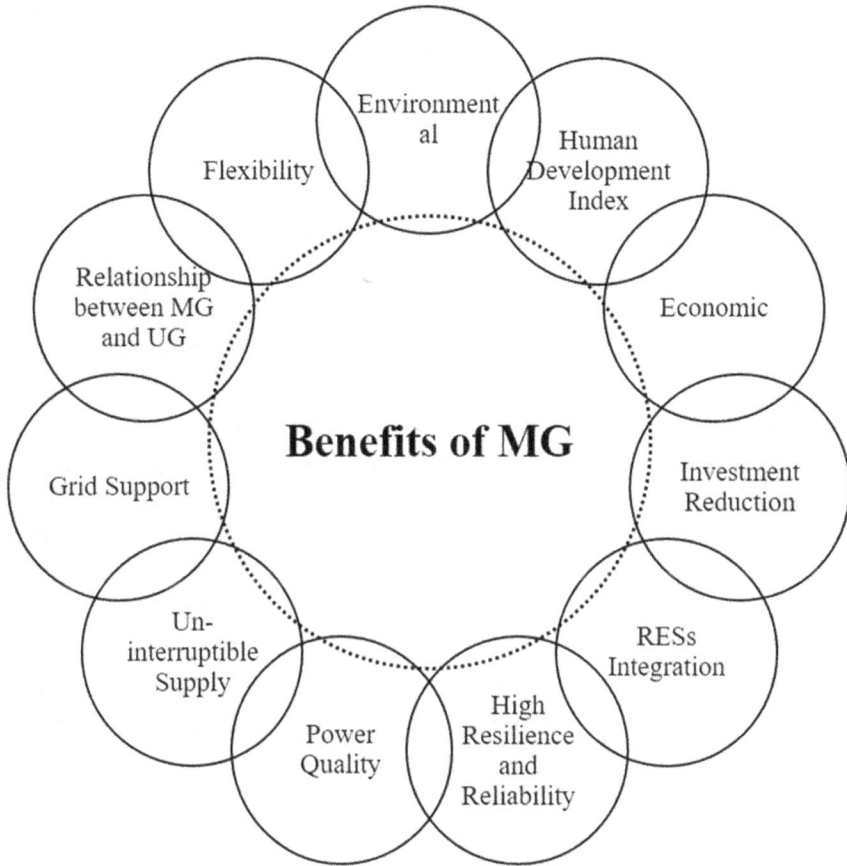

FIGURE 2.1   Benefits of MG.

2.1.2 Human Development Index Benefits

The availability of energy access improves the Human Development Index (HDI) in many ways, such as Refs. [3,4] (a) the availability of energy in the health and medical sectors in the developing economic nations will surely rise the HDI due to the presence of well-quipped hospitals, (b) the availability of energy for lighting and cooking purpose can improve the air quality of the house, thus providing a healthy environment to the people, (c) energy access to the farmers of rural agricultural areas where human and animal power is used for farming and yield purposes will significantly increase the HDI, (d) the nations with surplus energy generation help them to achieve the expected development goals that in turn leads to support the nation's economy, (e) for the people of rural areas, when they have access to electricity they are more likely to complete their education in a better manner, which in turn leads them to be employed and thus have the possibility to earn more money, (f) economical and reliable electrical energy helps and motivates entrepreneurs to start and grow their businesses, which may lead to an increase in the well-being and economic development of society, and (g) the availability of energy in society helps in removing hunger because of better processing and storage units that reduce food losses.

2.1.3 Economic Benefit

Depending on the initiatives and laws of the local market, MGs can be involved in DR markets, lower peak load prices by providing incentives, and also offer frequency regulating services to the UG. Moreover, MGs also provide protection against fluctuations in electrical bills. Moreover, the PnP capability of the MG may contribute to keeping the energy price low in the power market.

2.1.4 Investment Reduction

The MGs are usually installed close to the load, and the energy generation is typically distributed; therefore, the investment related to the expansion of transmission and distribution lines is considerably reduced.

2.1.5 RESs Integration

Globally, the ongoing energy crisis has created unparalleled motivation for RESs. In the last few years, to cope with the energy crises, RESs have contributed more than conventional sources; as a result, it is anticipated that in the upcoming five years, the RESs generating units will surpass the previous expectations. According to International Energy Agency forecast, renewable energy is predicted to grow by around 2400 GW in the upcoming five years, which shows an acceleration of 85% compared to the growth seen in the previous five years [5]. Moreover, these RESs are generally installed in a distributed manner; therefore, the MGs are becoming very important for gaining the benefits of RESs.

2.1.6 High Resilience and Reliability

In case of a power outage, the MG has the capability to continuously supply uninterruptible power to its end-users. The islanded capability of the MG can also be used to isolate the faults by separating the distribution feeders from the system.

2.1.7 Power Quality

The sensitive equipment in health care, laboratories, sophisticated manufacturing, and other sectors may require a high power quality that the conventional power system may not be able to provide. However, the MGs can provide enhanced power quality due to a better balance of load-demand graph, decentralization, improved power system restoration, and a reduction in power outages.

2.1.8 Uninterruptible Supply

Although the electrical systems in developed countries are generally stable, any power outage can be hazardous and costly. Cyber and physical attacks, ageing, and extreme weather conditions are all potential risks to the electrical infrastructure of the nation [6]. Hence, to cope with these risks, MG is a good solution as it has the capability to operate in IS mode. Therefore, in case of a power outage in the UG, the MG will isolate itself from the UG and use the on-site generation unit to ensure an uninterruptible and constant power supply.

2.1.9 Grid Support

MGs reduce grid congestion and peak loads. Besides these, MGs also offer numerous other grid services, such as energy and capacity. Moreover, when MG is operating in GC mode, it also provides ancillary services such as reactive power compensation, demand-supply balancing, and voltage regulation.

2.1.10 Relationship between MG and UG

MGs are considered as the main building component of the SG. Therefore, it is assumed that the future UGs may be the collection of different interlinked MGs that appropriately manage the demand and supply at the micro level.

2.1.11 Flexibility

MGs can be designed to meet the energy demand of a specific community and can be extended easily as needed.

2.2 CHALLENGES IN MG

Even though MG offers numerous benefits but during its development, there are numerous challenges that need to be discussed. The challenges in the MGs are broadly divided into technical, regulatory, market, and economic challenges, as presented in Figure 2.2.

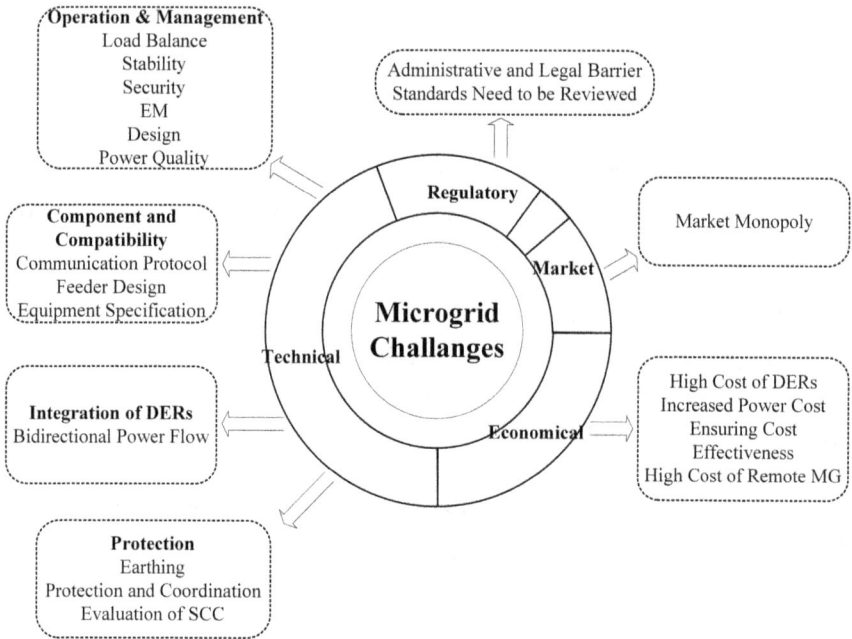

FIGURE 2.2   Challenges of MG.

2.2.1 Technical Challenges

The technical challenges in an MG are broadly classified into four categories, i.e., operation, component and compatibility, integration of DGs, and protection. These challenges are explained below in detail.

2.2.1.1 Operation and Management

An MG experiences the following operational and management challenges:

- **Balance between Load and Generation in IS Mode:** One of the most significant challenges that needs to be addressed is the graph balancing between load and generation. A significant or sudden load variation in IS MG can lead to system instability.

- **Stability:** The stability of an MG plays an important role in providing a smooth and uninterruptible power supply to end-users. The instability in an MG usually occurs due to transient events that need to be monitored, forecasted, and estimated very critically. As MG is a combination of different components such as energy generating resources, loads, virtual synchronous generators, storage devices, and PE converters, it is therefore very challenging to maintain the system's stability [7]. Moreover, the interment and stochastic nature of RESs-based DG units can further increase their complexity to maintain system stability. Therefore, MGs' designers and manufacturers should conduct a detailed comprehensive study to ensure the stable operation of the system [8].

- **Security of System:** To maintain the security of the system in an MG, emergency actions and contingency planning should be followed strictly. The actions under emergency operations include load shedding, demand side management, unit shutdown, and islanding. The actions under contingency planning include economical rescheduling of the generation to facilitate the system load and voltage/frequency of the end-user.

- **Energy Management (EM):** Generally, energy regulation in an MG involves the fine-tuning of several parameters that are used to find the best solution. Due to unspecific instances of varied ambiguity, it is very difficult to find the ideal value of these parameters in an MG [9].

- **Design:** The planning requirements and the design modelling of MG, especially RESs-based MG are different from the conventional resources-based system, which most of people are unaware of Ref. [10]. Hence, the unawareness of the people results in a poor MG design that reduces its longevity. Therefore, it is important to have a full understanding and knowledge of the available energy resources and demands of end-users while designing an MG. Moreover, the effect of varying energy generating resources and demand on the reliable energy supply should also be considered in the MG design process.

- **Power Quality:** The quality of active and reactive power of an MG is greatly and strictly affected due to the presence of non-linear loads, DG units, power conversion devices, and switching devices. Therefore, to maintain high power quality and ensure system stability, advanced controllers are very important and highly desirable. In an IS mode, the VP of an MG is not supported by the UG; therefore, it must have the capability to perform under unbalanced components and non-linear loads and effectively participate in power sharing mechanisms. If a proper control strategy is not designed, then numerous issues arise such as voltage deviation, non-proportional load sharing among the DGs, and harmonic distortion [11].

- **IS Mode Start-up Issues:** During the initial stages of an IS mode, the voltage and frequency of the system can be greatly affected by the high current intake. Hence, it may cause to trip and shut down the generators during the initiation phase. Hence, to address this challenge, an investigation and critical analysis of energy generating methods are needed. Moreover, the development and design of an intelligent and robust controller is required that is specialized in MG operations.

- **Identifying the MG Operation:** As an MG consists of numerous generators and loads that have different and varying operational behaviours and natures. Therefore, it is important to identify every integrated power generator versus load scenario and specify the situation, such as emergency shedding or temporary switching [12].

- **Supervisory Control and Data Acquisition (SCADA):** The control architecture of an MG should incorporate SCADA-based metering, protection, and control capabilities. Moreover, different provisions, such as state estimation functions, should be established for the system's diagnosis.

- **Smart Consumer:** Smart consumers play an important role in balancing the mismatch between demand and supply. These users are mainly interested in the lower cost of electricity or at least support the current comfort levels, simplicity, and accessibility. Currently, due to the active participation of residential consumers in demand management, the usage of information and communication technologies for information exchange has become more ubiquitous [13]. In the near future, an EMS will become an important part of smart homes to reduce utility bills, optimize energy usage, and meet the ever-increasing supply challenges while attaining the required level of comfort for its users [14].

- **Control System Analysis:** To ensure the high quality output waveforms and stability of an MG, a good design of a control system plays a very important role. While designing a control system, the configuration and operating modes of an MG should be studied in detail so that the designed control system can perform all the functionalities effectively. Moreover, the control system should be intelligent and flexible enough to incorporate different control strategies [15].

- **Load Flow:** It is important to analyse the load flow in every MG configuration and operating condition to determine the voltage levels and direction of current flow. It is a very challenging task to list the related loads and determine their specific values. Moreover, the loads with variable profiles may increase the complexity.

- **Inertia:** The integration of RESs-based DGs in an MG not only causes some degree of intermittency and uncertainty but also causes the issue of low inertia. This issue further increases the operation and control of an MG and significantly affects the frequency of the system, which may lead to system instability. In traditional power systems, rotational inertia is related to the rate of change of frequency and minimum frequency; however, the frequency deviation in an MG is inevitable. To overcome this challenge, a robust control scheme should be developed that mimics the behaviour of a conventional synchronous generator to provide synthetic inertia to the system.

2.2.1.2 Component and Compatibility

An MG experiences the following component and compatibility challenges:

- **Communication Network:** Communication, measurement, and information technologies play a very important role in the efficient performance of the MG system. The CN greatly affects the operational mode, protection schemes, topology, EM, power management, etc. Therefore, proper CN is required to monitor, stabilize, and analyse the MG at different levels of hierarchy. Therefore, the communication topology used in an MG must be reliable, cost-effective, secure, have fewer repetitions, have a high bandwidth, and have a good transmittable range. The conventional wired communication technologies were generally used due to their high security and reliability, but due to the high penetration of DGs, these communication topologies became very expensive due to the high complexity of the MG system. Hence, to cope with this challenge, in recent years, wireless technologies have been used in MG systems to provide reliable and decentralized communication [16].

Likewise, some parameters such as voltage magnitude, power, frequency, state of charge of ESS, and phase angle should be controlled and monitored in MG. Therefore, a reliable and efficient communication topology and control architecture are required to locally update the information. The wired communication technology can be used for data collection and transmission, but it may cause system instability in large systems with numerous DGs and multiple users. Due to these shortcomings, wireless technologies are widely used in an MG due to their high flexibility, low installation cost, and suitability for urban areas. Hence, the proliferation of highly penetrated RESs-based MG is greatly improved by using wireless communication technology for data collection, monitoring, and control [11].

- **Feeder Design:** In MG, feeders are designed based on energy generation and demand, just like a conventional power system. However, it will become a challenge in an MG when the demand for the feeders remains unfulfilled [17].

- **Specifications of Equipment:** The single-line diagrams and specifications of all the equipment in the power system play a very important role in the planning and designing process. Therefore, the basics for demonstrating and illuminating the operating behaviour of MG are important [18].

2.2.1.3 Integration of DGs

An MG experience two major challenges when DGs are integrated with the system. These are;

- **Voltage Variation:** Due to continuous increase in the integration of DGs in a system, it causes a voltage rise, which is one of the biggest challenges that needs to be addressed. In the case of a solar PV system, due to its intermittency, it causes a short-term variation in voltage that results in disruptions in the protection system and power regulation, which shortens the life of the equipment [19].

- **Bidirectional Power Flow:** When a DG with a low voltage level is connected to the MG system, it may cause a reverse power flow. The un-intentional reverse power flow poses challenges with voltage control, power flow patterns, protection coordination, current distribution, etc. [20].

2.2.1.4 Protection

As an MG has the ability to operate in IS mode as well as in GC mode it is very important to protect the MG in both of these operating modes in case of a fault. The traditional protection schemes such as thermomagnetic switches and fuses are not very feasible in this scenario. Instead, MG protection brings numerous technical challenges such as:

- **Structural Changes:** A main challenge concerning the protection of MGs is due to internal faults, specifically when an MG is operating in IS mode. Hence, any changes in the MG configuration have a crucial effect on the short-circuit current. Therefore, a suitable and reliable protection scheme is very important for internal faults as well as for faults in the MV distribution grid. The occurrence of any internal or external fault (MG operating in GC mode) results in high short-circuit currents that must be detected quickly by the protection scheme to ensure system stability [21]. For instance, when an MG is operating in IS mode, and if any fault occurs then there will be a significant reduction in short-circuit current, particularly in the absence of synchronous generators. As the PE inverter has a very limited capacity to feed short-circuit current, it will affect the setting of the protection components. Similarly, in case of any external fault, the protection system of the MV grid needs to separate the MG from the MV system as soon as possible by sending opening the signal to the circuit breaker. However, if the protection system of the MV system did not respond properly, then the MG protection scheme should send the opening command to the circuit breaker [22].

- **Earthing:** In terms of MG protection, neutral earthing is a difficult and vital issue to address. The main reason behind this difficulty is the use of different power sources such as converters and spinning generators, PnP phenomena in MG, and integration with UG. Moreover, the installation of neutral earthing is mostly done by the local grid regulatory authorities, but the maintenance and distribution of earthing may cause certain issues [22].

- **Change in MG Topology:** The topology of the MG can be changed at any time due to the intermittent nature of the RESs-based DG units (such as wind turbines and PV), ESS, and loads (available or not, ON or OFF, due to various reasons such as maintenance, loss minimization, etc.). Therefore, the protection scheme should be designed considering all the possible operating conditions. However, considering different operating conditions of an MG, a protection scheme with a single group of settings is not efficient. Therefore, adaptive protection schemes should be designed to ensure system protection in different operating conditions [23].

2.2.2 Regulatory Challenges
2.2.2.1 Standards
In contrast to conventional power systems, an MG is comparatively a new industry; therefore, protocols and standards for DG integration, participation in traditional deregulated power markets, safety, and protection should be reviewed. To ensure a safe and proper integration of MGs with the UG standards, such as IEEE 1547 and G59/1, should be revised and reconstructed. Moreover, the researchers also needed to focus on the review of IEEE 2030.7-2017-IEEE standards that were developed for the MG controllers [24].

2.2.2.2 Legal and Administrative Barrier
One of the biggest challenges in attaining an MG regulatory is the lack of explanation of clear legal identification for MG. Due to a lack of explanation, legal issues arise due to uncertainties concerning whether the MG is a separate entity or assumed to be a distribution utility, and whether the existing legal framework is enough to administer the sale, distribution, generation, and purchase of MG electrical energy [25]. Moreover, the current legal framework also lacks a definition of the standards for connecting DGs with UG; as a result, the integration requirements greatly vary from utility to utility. Although IEEE standards 1547.4 cover communication, control, protection, safety considerations, power quality, MG functionality, and VP specifications but still a lot of legal explanations and standardizations are still needed [26].

2.2.3 Market Challenges
If the MGs are allowed to allocate energy independently based on load priority in case of any major grid disruption, then a key question arises: Who will be in charge of determining the energy prices during the outage? When the UG is disconnected from the MG, the centralized electrical market also loses control over the energy prices generated by the MG. Thus, as a result, a MG may take advantage of the market monopoly and may start selling electricity at a high price. Therefore, to support long-term MG development and avoid such monopolies, a well-structured electricity market should be implemented and established.

2.2.4 Economic Challenges

Although the MG has numerous positive economic impacts, such as an improved job market, reduced energy cost, and an enhanced security system. However, there are also some negative economic aspects, such as the high maintenance and installation costs, high upfront infrastructure costs which are less feasible for communities with fewer financial resources. Therefore, in this subsection, different economic challenges of the MG are discussed below in detail.

2.2.4.1 High Investment

Some of the components of MG, such as ESS, fuel cells, and grid management software, are still not available commercially. As a result, all these increase the cost of the MG; therefore, for communities with limited financial assistance, the MG projects are not very appealing.

2.2.4.2 DGs Cost

In economic prospective, the main challenge of an MG is its high installation cost. However, to overcome this hurdle, the government should provide some subsidies in the MG development sector to promote investors. For the sake of achieving the global goal of increasing RESs-based energy generating resources and cutting carbon emissions by 50% by 2050, the government should at least provide subsidies temporarily.

2.2.4.3 Economic Operation

To accomplish a cost-effective operation of the MG, all the operations, such as economic load dispatch, energy generation scheduling, and efficient power flow should be used.

2.2.4.4 Generation Cost

As compared to AC and DC MGs, the architecture of the HMG system is more complex. Hence, the complex architecture may lead to the possibility that the electricity generation cost will increase in these systems.

2.2.4.5 Cost of Remote MG

The remote MGs are usually installed in remote or rural areas that are far away from urban areas. Therefore, installing a remote MG may be more expensive due to the high transportation and increased maintenance costs.

2.2.4.6 Fixed Cost Compensation

The MG offers some consumer-friendly functionality such as end-users are encouraged to generate electricity on their own, then they will purchase less electricity from the utility. Some end-users may generate more electricity than they need, in that scenario, they can sell the additional electricity to the utility. An end-user who uses a net-metering system can get benefit from this functionality and remain connected to the UG without paying any extra expenses. On the contrary, end-users who do not use the net-metering functionality may face the expenditures of infrastructure expansion because infrastructure extension expenses are linked with the utility tariffs [27].

2.3 TECHNICAL ASPECTS OF MG

There are some technical aspects that should be considered while integrating an MG with the power system. The different technical aspects that are discussed here are presented in Figure 2.3.

2.3.1 Power Quality

It is the analysis, measurement, and enhancement of the bus voltage to maintain the high quality sinusoidal waveform according to its nominal frequency and voltage [28]. It is the most important component when an

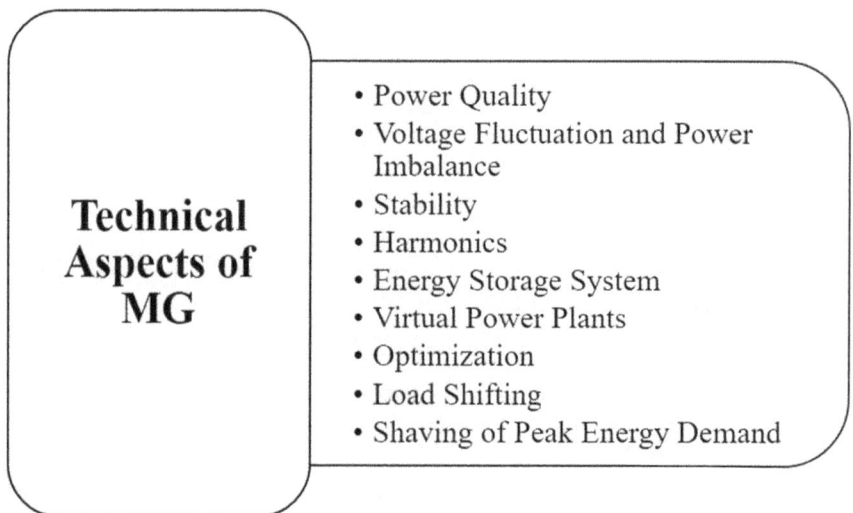

FIGURE 2.3 Technical aspects of MG.

MG is operating in GC mode and must be handled carefully. As most of the distributed energy generating resources in an MG are based on RESs and are usually connected to the MG through the PE interface therefore the high usage of the switches causes power quality issues. The integration of different intermittent RESs-based DG units such as PV and wind, also affects the power quality of an MG. Moreover, another factor that affects the power quality of the system is the non-linear load. So, it is very important to implement advanced intelligent control schemes that mitigate the negative effects discussed above and provide high quality waveforms at the output. Poor power quality also affects the power quality-sensitive components, and in the future, poor power quality may result in poor electricity prices [29].

2.3.2 Voltage Fluctuation and Power Imbalance

When an MG is disconnected from the UG, it causes voltage fluctuation and power imbalance in the system. These effects are mainly due to the zero or low inertia and slow dynamic response of the RESs-based DG units. To cope with the aforementioned challenges, mainly control schemes are designed along the Flexible AC Transmission System (FACTS), Static Synchronous Compensator (STATCOM), and storage devices are used to ensure the smooth transition of MG from GC mode to IS mode. To improve the dynamic performance of the MG, the authors in Ref. [30] designed a distributed STATCOM with a PID controller that is optimized by the grasshopper algorithm. Similarly, the authors in Ref. [31] proposed a distributed STATCOM device with FACTS technology to improve the operational behaviour of an MG. Moreover, the authors also presented the relationship between MG's operational behaviour and the FACTS devices, and from the simulation results, it can be concluded that the proposed method effectively and efficiently supplies the reactive power to the system.

2.3.3 Stability

Due to different operational characteristics of DGs and MGs, the issue of the stability of the system arises. There are three core reasons due to which stability issues arise in the system, these are [32]: (a) DG units have low inertia that results in frequency and angular instability, (b) fewer oscillations in the frequency due to a shift in power sharing ratio, and (c) low voltage stability due to reduced energy distribution. However, the stability and power quality of the system can be improved by maintaining a balance between the demand-supply graph and by decentralizing the power supply [29].

2.3.4 Harmonics

To achieve high reliability, the optimization of MG operating in both IS and GC modes is very important. In both of these operating modes, the harmonics pose a threat to the reliable operation of the system if they are handled with precautions [33]. In case of DC MG system, the charging and discharging of the capacitors cause an increment in the harmonics. Therefore, if the harmonics are not taken seriously, they may jeopardize the ESS because of its highly responsive nature to be affected by the harmonics [34]. Moreover, the PE components (especially the switching devices) are the main cause of the introduction of harmonics in the power system. In the literature, numerous harmonic mitigation techniques were developed and categorized as active and passive filters. Active filters are used to eliminate the harmonics of any order, while passive filters are used to eliminate the high-order harmonics.

2.3.5 Energy Storage System

Due to the dependency of RESs on the environment, they cannot independently supply uninterruptable power to end-users. Therefore, to meet the energy demands and ensure a smooth power supply, these RESs-based DGs can be used along with the diesel engine-based DGs. Although this combination may meet the energy demands but the diesel engine-based energy generator is not eco-friendly, so it is not a very feasible solution. Therefore, in recent years, the combined use of RESs-based DGs with the ESS units has significantly increased. The ESS units store energy when the generation is high compared to consumption and use the stored energy during low energy generating hours. Besides meeting the energy demand during off-peak generation hours, there are numerous other benefits of ESS units, such as increasing the power factor of the system, achieving power quality and system stability, reducing oscillations, and achieving a balance between demand-supply [35].

2.3.6 Virtual Power Plants

Compared to the installation of large power plants, RESs are usually installed in a distributed manner close to the end-users. As the RESs-based DGs have intermittent power generation and have low inertia; hence to cope with these challenges, the concept of Virtual Power Plant (VPP) was proposed in 1997 [36]. The concept of implementing this idea is to group different DGs that include, e.g., RES, ESS, distributed power plants, EVs, and controllable loads. DGs can be distributed over wide geographic areas within a given VPP and the DGs are then managed by EMS [37]. Fundamentally, the operation of this EMS can be described in three steps, i.e., (a) in an input, it receives the data of the actual energy production and the signal from the market at definite time intervals, (b) in a second step, the EMS forecasts the output power of the RESs-based DG units and the demand of the load, and (c) it coordinates different interconnected RESs-based DGs to maximize its objective function [38].

When viewed from an external point, VPPs act as conventional transmission-connected plants; therefore, VPPs can be characterized by different factors such as voltage regulation capability, ramp rates, and scheduled output [39]. Practical examples of VPPs are developed in the UK and Germany, where multiple units are integrated together. In the end, it can be concluded that the VPPs enable the DGs to trade in energy markets and deliver services to facilitate transmission system management.

2.3.7 Optimization

Optimization is generally related to the placement of DG and network reinforcement in planning phase, and the dispatch of DG and network configuration in operation phase. In the literature, numerous different optimization objectives have been presented, such as load balancing, reducing network losses, maximizing consumption of RESs, improving voltage profiles, minimizing fault currents, and reducing service interruptions. To achieve these objectives, different algorithms have been designed based on linear programming, fuzzy approaches, simulated annealing, genetic algorithms, heuristic techniques, etc. To perform the active distribution network reconfiguration, real-time network data is required, which can be provided by an advanced metering system that is supported by a proper and reliable data management infrastructure [40]. Another aspect of optimization in the operation phase is the dispatch of DGs into the MGs. By using an optimization technique for DGs, dispatching significantly improves and increases the acceptance of DGs in the distribution system [41]. The aspects of active distribution network optimization in the planning phase include network configuration in non-real time, wiring, and protection device installation. It should be noted that during the optimization of the planning stage, it is important to take the uncertainties of the generations and loads under consideration [42].

2.3.8 Load Shifting

Load shifting is the process of gradually shifting the energy demand during peak hours into periods when the energy demand is low. It maximizes the usage of the current generation resources while attenuating the minimum and maximum of the energy consumption curve. Load shifting can be attained by direct control of electric components (usage of the appliances is not restricted to a specific duration of day) or by adopting policies that encourage the users to use energy in specific time slots [43]. In case of self-generation from the RESs, load shifting encourages the users to use more energy during peak generation time to get economic benefits and use less energy when the generation is low or during peak demand hours. Besides these benefits, load shifting also helps to reduce congestion problems in the UG [44].

2.3.9 Shaving of Peak Energy Demand

To avoid voltage fluctuation, grid failure, and grid instability, the energy generation must not be less than the energy consumption at any given instant. Hence, to meet the high load demands, instant energy generating units (such as diesel generators) can be used. These expensive energy generating units are turned on during peak demand situations to ensure energy availability and turned off during off-peak energy demand. In the literature, the use of BSS and the application of DR techniques can also be used for peak demand shaving. BSSs are used in residential [45], industrial [46], and tertiary sector consumers [47]. DR as a management system can also be used to reduce the peak load [48].

2.3.10 Communication in MG

A robust and reliable communication system plays an important role in the optimal and stable operation of the MG. A communication system in an MG is considered a bridge between the physical system and its protection and control processes. However, the presence of loads, DGs, and communication between the nodes significantly increases the complexity of the control and operation of the MG; thus, as a result, the communication complexity also increases. Therefore, it is very important to choose an appropriate communication system that offers a high level of security, reliability, and performance with bidirectional connectivity between the resources of the MG and its connectivity with the power system. To select an appropriate communication system, the type of CN and Communication

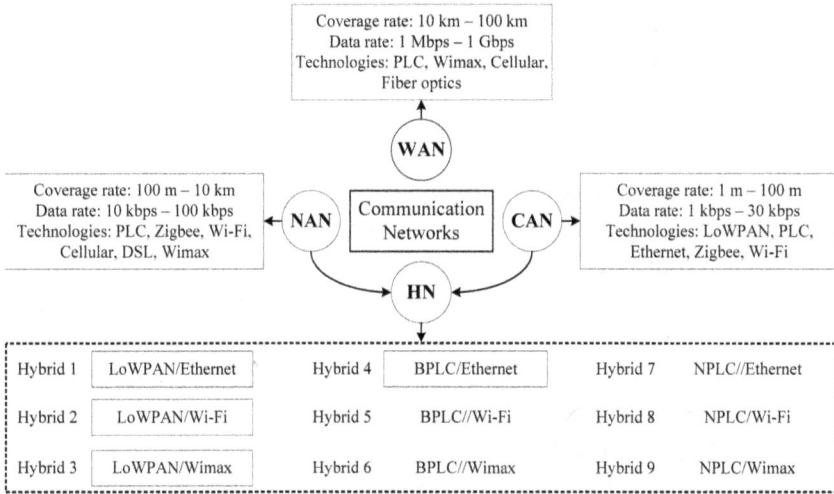

Coverage rate: 10 km – 100 km
Data rate: 1 Mbps – 1 Gbps
Technologies: PLC, Wimax, Cellular, Fiber optics

WAN

Coverage rate: 100 m – 10 km
Data rate: 10 kbps – 100 kbps
Technologies: PLC, Zigbee, Wi-Fi, Cellular, DSL, Wimax

NAN

Communication Networks

CAN

Coverage rate: 1 m – 100 m
Data rate: 1 kbps – 30 kbps
Technologies: LoWPAN, PLC, Ethernet, Zigbee, Wi-Fi

HN

Hybrid 1	LoWPAN/Ethernet	Hybrid 4	BPLC/Ethernet	Hybrid 7	NPLC//Ethernet
Hybrid 2	LoWPAN/Wi-Fi	Hybrid 5	BPLC//Wi-Fi	Hybrid 8	NPLC/Wi-Fi
Hybrid 3	LoWPAN/Wimax	Hybrid 6	BPLC//Wimax	Hybrid 9	NPLC/Wimax

FIGURE 2.4   Classification of communication network.

Technology (CT) plays an important role. In this subsection, the CN and CT (wired and wireless) along with their different sub-types and different communication protocols are discussed in detail.

2.3.10.1 Communication Networks

Based on the application requirement, data rate, technology, coverage area, and functions, the CNs are classified into four categories, as presented in Figure 2.4. These are: (a) Customer Premises Area Network (CAN), (b) Neighborhood Area Network (NAN), (c) Wide Area Network (WAN), and (d) Hybrid Network (HN).

- **Customer Premises Area Network:** CAN corresponds to the first segment of the communication process and provides low bandwidth. Depending on the consumption profile, it can be further classified based on residential, commercial, and industrial applications as (a) Home Area Network (HAN), (b) Business Area Network (BAN), and (c) Industrial Area Network (IAN). CAN provide two-way communication between the user accumulator and the smart appliances. At this level, real-time data (frequency, current, power, and voltage values) is collected and used in DR and DSM. The Data Transfer Rate (DTR) CAN is usually up to 30 kbps with a coverage area of 1–100 metres. Different technologies are used by CAN, such as Zigbee, Wi-Fi, Low Powered Wireless Personal Area Network

(LoWPAN), and Power Line Communication (PLC) (i.e., broad-band and narrow-band PLC) [49,50].

- **Neighborhood Area Network:** NAN provides bidirectional communication between the user accumulators and the MG controller. The data collected by the smart metres is sent to the controller to perform its mandatory control actions. In a centralized control architecture, a central controller of the MG communicates with the local controllers of the DERs to achieve different control objectives. In a decentralized control architecture, the local controllers of the DERs also communicate with each other to accomplish the control objectives [51]. Compared to CAN, this network configuration has a high coverage area, i.e., 100 m–10 km, and a DTR of usually 10–100 kbps. The different technologies used by NAN are mesh network, Wi-Max, Digital Subscriber Line (DSL), Zigbee, PLC, cellular, and Wi-Fi [52].

- **Wide Area Network:** WAN, also known as Access Area Network (AAN), provides bidirectional communication over long distances, typically ranging from 10 to 100 km. WAN supports numerous functions (such as sensing and monitoring) between different grid headquarters and substations and also provides communication between the MG and the UG. Moreover, it also has the capability to exchange information with external users such as Market Operator (MO), Distribution Network Operator (DNO), or other MGs [53]. WANs require high bandwidth to handle the huge amount of data that is generated by external users and power grids; therefore, the range of DTR is typically from 1 Mbps to 1 Gbps. A high-speed optical fibre communication medium is preferably used in WANs [54]. However, more options could be PLC, Wi-Max, and cellular [55].

- **Hybrid Network:** HN is a combination of different HANs and NANs technologies and comprises nine different architectures [54]. These are (the first technology is HAN and the second technology is NAN): (a) LoWPAN/ethernet cable, (b) LoWPAN/Wi-Fi, (c) LoWPAN/Wi-Max, (d) BPLC/ethernet cable, (e) BPLC/Wi-Fi, (f) BPLC/Wi-Max, (g) NPLC/ethernet cable, (h) NPLC/Wi-Fi, and (i) NPLC/Wi-Max.

2.3.10.2 Wired Communication Technologies

Wired communication technologies (WDCTs) are used in different applications such as demand side networks and automatic metre

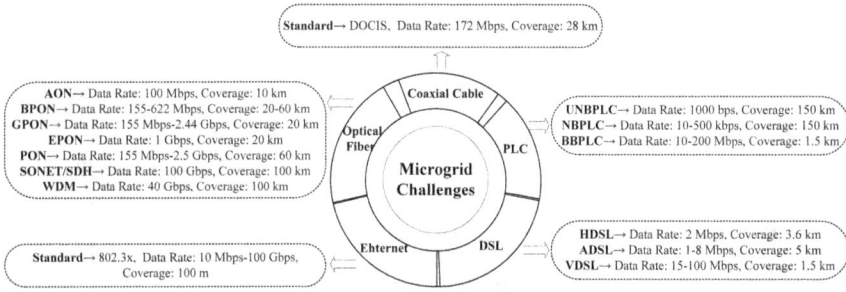

FIGURE 2.5 Wired communication technologies in MG.

reading [56]. Moreover, WDCTs are also a preferable option for the already existing utilities. Its high structural flexibility and readily existing infrastructure allow bidirectional communication inside and outside the MG. Different technologies fall under this category, such as PLC, Ethernet, coaxial cable, and optic fibre, that are discussed below in detail. Moreover, the DTR and the coverage of different WDCTs are presented in Figure 2.5.

- **Power Line Communication:** PLCs are widely used in different electric power applications such as load control, distribution network services (remote metering), and tele-protection between substations. Moreover, PLCs are also used to exchange information in the already installed MV and LV distribution networks by modulating carrier signals. With the advancement in technology, PLCs are further divided into three types, i.e., Narrow-Band PLC (NBPLC), Broad-Band PLC (BBPLC), Ultra Narrow-Band PLC (UNBPLC), and among them, NBPLC and BBPLC are commonly used. The main difference between these two commonly used PLC technologies is the operating frequency, i.e., BBPLC operates at high frequencies; on the contrary, NBPLC operates at low frequencies, as presented in Figure 2.5 [56]. Based on the technology used, the distance limits and data rates also vary accordingly, i.e., low-frequency carriers (100–200 kHz) impressed on HV transmission lines may cover long distances with a DTR of a few hundred bps, carrying control signals, analogue voices, and telemetry. On the contrary, to achieve high DTR, the coverage area must be reduced. For example, a Local Area Network (LAN) operating at millions of bps may be able to cover only one floor of the building.

The applications of PLC are not only limited to the power system but they can also be used in offices and homes to interconnect appliances that have an Ethernet port. In PLC-networked offices and homes, it provides internet access through a BBPLC modem through which any electric appliances can be connected. The adapters that provide such connectivity are often known as Ethernet over Power (EOP) adapters. These adapters allow the sharing of data between the devices without using dedicated network cables. Moreover, with the advancement in technology, a new mechanism called E-Line has been discovered that is able to support high DTR (up to 1 Gbps) and has operating bands between 100 MHz and 10 GHz [57].

Besides these numerous applications and advantages, the PLC suffers from different noise, distortion, and attenuation problems [58]. This is mainly because of the design of the distribution system, which is mainly designed for 50/60 Hz transmissions; therefore, it has a limited capability to carry high frequencies. As a result, it may cause problems related to communication and may result in inappropriate operation of the controlled components [57]. Besides these limitations, PLCs are considered to be more secure as compared to Wireless Communication Technologies (WSCTs) in terms of cyber-attacks [59].

- **Optical Fibre Communication:** The main advantages of this technology are its high data-carrying capability over long distances with extremely low losses. As compared to the copper wires, optical fibres allow long distance links among the repeaters and amplifiers with a high data volume. Moreover, another advantageous feature of optical fibre is that it does not have cross-correlation effects even if multiple cables are running alongside each other for large distances. Moreover, it has the capability to perform efficiently in high Electromagnetic Interference (EMI) areas such as railroad tracks, power lines, and utility lines. Due to these advantageous features, it is used by numerous communication firms and companies to efficiently transmit television, internet, and telephone signals. The optical fibre technology can be used in MG applications, especially in WANs, to exchange real-time data and energy market information between the MG controllers and DNOs [60]. Besides these numerous applications and advantages, the main disadvantage of this technology is its expensive and complex operation and installation. Due to these limitations, optical fibre systems are most favourable to install in long distance applications, where they can be used to their full capacity to compensate for the expenses of installation.

- **Ethernet:** It belongs to a family of computer networking and is based on IEEE 802.3 standards for LANs and Metropolitan Area Networks (MANs). Compared to the initial design, i.e., 802.3 protocol, a lot of modifications (like clock synchronization, new priority schemes, etc.) are made to this technology to make it applicable for real-time communication to fulfil the requirements of industrial applications. Due to recent advancements, the DTR and link distances have significantly increased. Therefore, the DTR can now even reach 100 Gbps in the Gigabit Ethernet. Beside high DTR, this technology is easy to install, easily available, and has cheap integrating circuits. Due to these reasons, they are widely used in computational sectors, and most of the computerized devices already have a built-in Ethernet interface that shows its dominancy in the entire communication infrastructure [61].

  The usage of Ethernet is now also extended to SCADA systems; most of the standard protocols, such as DNP3 and Profibus, and vendor-specified protocols such as Modicon and Modbus offer a completely new interface to operate over Transmission Control Protocol/ Internet Protocol (TCP/IP). With these new operating features, copper wires can be used to achieve the connection (links). However, in sites where the level of EMI is high, such as railway applications and power substations, synchronous optical networking can be utilized to achieve the connection.

  For security reasons, it is a good practice to avoid the use of the internet in SCADA communication. However, despite the security risks, most of the communications in Human Machine Interface (HMI)-based SCADA systems are done through Ethernet TCP/IP [62]. The use of SCADA in MG applications may be a natural option to control and monitor ESS, heat generation, electric power generation, and numerous other ancillary services. Therefore, Ethernet is widely used in MG applications where it performs point-to-multi-point or point-to-point connections like links between DGs and MG controllers or links between DGs.

  Beside the above-discussed wired technologies, coaxial cables and DSL can be used in MG applications. The main advantages of coaxial cables are their low cost and easy installation but they low bandwidth and high noise and attenuation. Similarly, DSL is also an economically viable solution and has easy installation, but it results in poor data quality that may affect the control functionality of the system [63]. Moreover, the coverage area of DSL is also very short as can be seen in Figure 2.5.

2.3.10.3 Wireless Communication Technologies

Generally, WDCTs are used for the exchange of information inside the MG architecture, such as primary controllers, PE converters, and measurement devices because of their close contact among themselves. However, the signals that flow among different kinds of equipment in an MG, such as tertiary controllers, SCs, and DGs, usually use PLCs and WSCTs. Different features like scalability, maintenance cost, protection against EMI, operation cost, data losses, etc., should be considered when selecting an appropriate CT according to the application.

Compared to WDCTs, WSCTs are more versatile due to their high reliability, scalability (simply adding the routers and nodes without using additional installation cables), and cost. Among the wireless technologies, cellular technologies such as 3G, 4G, and 5G are very robust but they need to be operated only at licensed radio frequencies. Hence, the use of cellular technology by third-party services causes an increase in operating costs [64]. On the contrary, technologies like Zigbee and Wi-Fi use unlicensed frequencies, which results in a better cost-benefit ratio. However, these technologies have high-security issues and line-of-sight requirements and are highly vulnerable to interference. Due to these technical limitations, it may result in data losses and high latency [65].

From the above discussion, it can be concluded that the performance of cellular networks is far better than that of Wi-Fi and Zigbee technologies. However, when selecting a network system, aside from performance, there are other matters that should also be considered, such as cost-benefit ratio and availability of resources. To clarify this point, the authors in Ref. [66] consider a scenario in which different networking features and devices, such as access points, routers, and firewalls, are used properly. In this scenario, the use of Wi-Fi networks for the exchange of information among the SCs may be more feasible as compared to cellular networks (cost-benefit ratio). Although the latency of Wi-Fi networks is about tens or hundreds of milliseconds but it is possible to attain the response of the SC in a few seconds. It should be noted that cellular networks are very feasible for the exchange of information between the SCs, but networks based on cellular technologies become very expensive when service agreements are well-defined to ensure quality requirements. Furthermore, in the future, it is expected to deploy the MGs in rural areas where there is a chance that cellular networks may not always be available. Due to these reasons, in this scenario, Wi-Fi technology is considered to be more feasible than cellular technologies.

Recently, in a group of cellular technologies, 5G has emerged as an emergent technology and has numerous applications in urban MGs, specifically those 5G networks that are based on Massive Machine Type Communication (mMTC) and Ultra-Low Latency and Reliable Communication (ULLRC) features. The nMTC standard was established for Internet of Things (IoT) industrial applications with the capability to connect a huge number of devices. Therefore, to preserve the reliability of the network, low DTRs are expected in this standard. Due to these features, nMTC is feasible for the tertiary control applications of the MG. Moreover, the 5G networks that are developed on the basis of ULLRC standard are feasible for the SC applications of the MG [67]. Although it has a wide range of applications and usage in urban areas, however, its installation and deployment in remote areas is less applicable and very expensive. If the high frequency band is employed, then to overcome the effects of line-of-sight obstacles and improve the coverage, numerous hotspots will be required to ensure high quality transmission. Thus, as a result, the operating and maintenance costs will increase. Similar to the above scenario, in this scenario, Wi-Fi has an advantage over 5G technology by considering that the low cost and private Wi-Fi networks are deployed and directional antennas are used to improve coverage [68].

Recently, a new WSCT known as LoWPAN has been considered as an alternative solution for SG connectivity and, in a few cases, also suggested for MG [69]. LoWPAN technologies such as NB-IoT, Sigfox, and LoRa provide wider coverage and lower operational and implementation costs compared to cellular technologies. It should be kept in mind that the existing literature studies only describe the possible implementation of LoWPAN technologies by evaluating their DTR, energy consumption, and coverage without precisely evaluating the suitability of this technology to accomplish the control requirements of MG [70].

CTs such as Bluetooth, Zigbee, and PLC show very good performance in indoor environments and have the ability to efficiently exchange information among devices, making them very appropriate for the tertiary control of MG. As the tertiary controller mainly focuses on the coordination and management of the MG; therefore, the latency and reliability requirements of the CN are lower as compared to the primary and secondary control levels. These technologies in the outdoor environment send the logged information from the devices to the tertiary control, which is then used as input for load management or load forecasting.

The most significant quality of wireless communication networks is their ability to change from infrastructure-based to ad hoc communication [71]. Moreover, it is not mandatory that the CNs be identical to the physical network of the power system. Furthermore, different communication topologies, such as star, bus, and hybrid, can be used in wireless networks to improve reliability, redundancy, and speed while avoiding traffic losses and congestion. Similar to the WDCTs, the coverage area of the WSCTs also depends on the type of technology. The area coverage of Bluetooth is limited to a few metres, whereas the coverage of Wi-Fi can be about 100 metres. Moreover, another issue related to wireless communication is its penetration capability. The technologies operating at higher frequency bands have a lower penetration, while the technologies operating at lower frequencies have a higher penetration and are more feasible for MG applications that are installed in obstructed environments, e.g., cities or mountains.

Furthermore, a detailed categorization of WSCTs along with their DTR and area coverage is presented in Figure 2.6. The advantages and disadvantages, along with the frequency bands of different wired and wireless technologies suggested for MG applications are presented in Table 2.1.

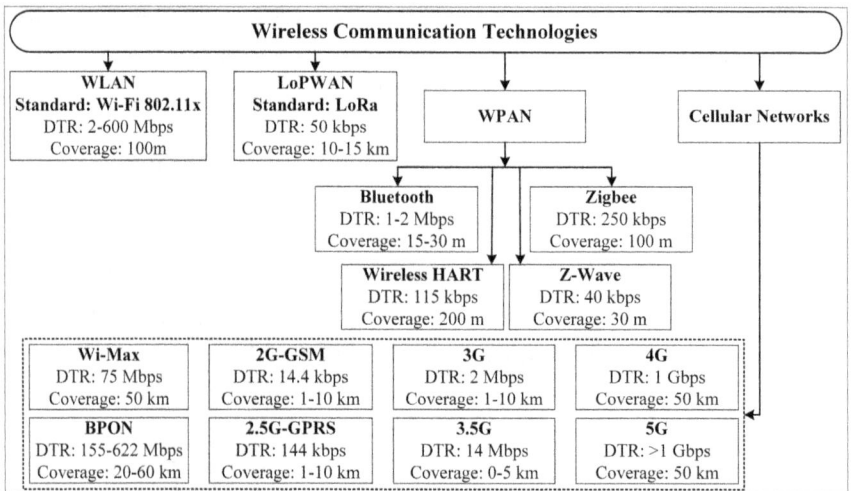

FIGURE 2.6   Wireless communication technologies in MG.

TABLE 2.1   Summary of Communication Technologies

Technology	Advantages	Disadvantages	Frequency Band
Wired Technologies			
PLC	Cost effective Home-plug compatibility Feasible for tertiary control	Non-flexible structure Complex scalability Noise and attenuation	<100 MHz
Optical fibre	Low latency Low interference High speed Feasible for primary and secondary control	Expensive maintenance Complex scalability Expensive technology	Visual spectrum and near infrared
Ethernet	Cheap integrating circuits High capacity Easy configuration Feasible for primary and secondary control	Complex scalability	<500 MHz
Coaxial cable	Low cost Easy installation	Low bandwidth High noise and attenuation	5–3000 MHz
DSL	Low cost Easy installation	Poor data quality High noise	25 kHz to 1.5 MHz
Wireless Technologies			
Cellular 3G	Third-party services Design for data packets transfer Feasible for tertiary control	Expensive QoS High operation cost	800/850/900/1800/1900/2100 MHz
Cellular 4G	Third party services Design for internet access Feasible for tertiary control	Expensive QoS High operation cost	2600 MHz
Cellular 5G	Low latency Design for internet access Feasible for secondary and tertiary control	New technology Non-wide facilities	mm-wave and sub-6 GHz
Wi-Fi	Easy scalability Easy configuration Cheap technology Feasible for secondary control	The increase in security causes increment in cost Susceptible to interference	2.4/5.9 GHz
Zigbee	Low power consumption Cheaper technology Feasible for tertiary control	Low speed	2.4 GHz
Bluetooth	Cheaper technology Feasible for tertiary control Low power consumption	Security issues Low penetration	2.4 GHz
LoRa	High penetration Feasible for tertiary control Low power consumption Easy scalability	Restricted downlink Slow speed	433/868/915 MHz

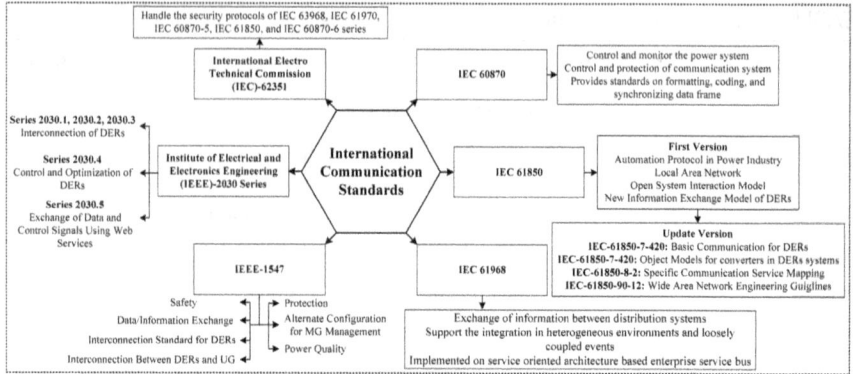

FIGURE 2.7 Communication standard deployed in MG.

2.3.10.4 Communication Protocols and Standards

To ensure a secure and reliable CN in an MG, it is important to strictly follow the standards and protocols of interconnectivity. Related to the concept of MG, the generation is usually distributed in nature, and the consumers are scattered over an area. As a result, a large amount of data is scattered in different places, so the standards and protocols related to communication must be followed. In this section, some standards related to communication in MG are briefly described and presented in Figure 2.7.

- IEEE Standard 2030: The IEEE 2030 standards define three different architectural perspectives: Information technology, communication topology, and power system [72]. From these three perspectives, it provides recommendations for MG compatibility and incorporates the standard framework for control and information sharing through CN in the power system. It also describes information related to logical connections, management of digital information, communications and lines, design tables, and data flows. Moreover, this standard identifies the interactions among functional domains of the UG from the view of each perspective and defines the relations among different domains, including the issues, constraints, and characteristics of data flow.

- IEEE Standard 1547: It describes the interconnection standards of DERs such as wind turbines, PV, and fuel cells with the power system [73]. It explains the guidelines to monitor, control, implement, and information exchange to support the business and technical operations of stakeholders and controllers of the DERs through direct communication. Moreover, the guide also includes templates for information exchange, information modelling, and the interface for information exchange.

- **IEC 62351:** It was developed by IEC Technical Committee (TC) 57 to handle the security of TC 57 series protocols, which include IEC 63968, IEC 61970, IEC 60870-5, IEC 61850, and IEC 60870-6 series [74]. Different security objectives, such as prevention of spoofing and playback, prevention of spying, transfer of data through digital signatures, intrusion detection, and ensuring authenticated access. These standards must be applied to the already existing security protocols, such as Transport Layer Security (TLS), that are widely used in many industrial and technical applications. A TLS provides many security services, like the protection of communication data and the mutual authentication of peers.

- **IEC 61850:** Electrical systems communications use the IEC 61850 framework as a foundation for information presentation [75]. Through this framework, all the concerned parties are able to share and efficiently use the information. This standard defines the naming agreements for data, the structure of the data confined to different devices, testing procedures, and how the devices are used by applications. There are various updated versions of this protocol that are presented in Figure 2.7 along with their major contribution to the communication system of MG.

- **IEC 60870:** These standards were developed by IEC TC 57 WG 03 for the control and monitoring of power systems, and the tele-control and tele-protection of related communications in power systems [76]. A fifth part of this standard describes the information related to sending the basic control messages between two power systems. Moreover, this standard is also feasible for different configurations: multi-dropped, star, point-to-point, etc. To be specified, it provides standards for formatting, coding, and synchronizing data frames of varying lengths.

- **2.4.4.6 IEC 61968:** These standards were developed by IEC TC 57 WG 14 and are related to the exchange of information between distribution systems [77]. IEC 61962 is designed to support integration that is based on heterogeneous environments and loosely coupled events with interface adapters and middleware services to achieve non-real message transactions. This standard is usually implemented on a service-oriented architecture-based enterprise service bus to linearize the interface numbers compared to conventional architecture (point-to-point), resulting in reduced maintenance and development costs.

2.4 CONCLUSION

Due to the high penetration of RESs into the power system, the popularity of implementing an MG system has significantly increased worldwide. Therefore, this chapter provides a detailed explanation of different benefits of an MG system, such as environmental, economic, and power quality. Moreover, besides numerous benefits, the MG system has numerous challenges and technicalities. Therefore, different challenges such as economic, regulatory and operation and technical aspects such as stability, load shifting, and power balancing are discussed in detail. Furthermore, as an MG requires a CN to perform its control actions, hence this chapter presents a detailed review of different wireless and wired CN technologies, topologies, and standards. It is also concluded that for the successful implementation of CN in an MG system, significant integration challenges can be faced because of the constraints of the protocols. These challenges can be overcome by relaxing the constraints of optimization and standardization [78,79]. Moreover, this chapter will help the researchers and engineers to clearly understand the advantages, challenges, technical aspects, and communication network.

2.5 EXERCISES

1. How the development of MG can improve the life style of community?

2. What measures should be taken that can further enhance the economic benefits of an MG system?

3. What are the other possible technical aspects that should be considered?

4. Communication is the core of component of MG, try to develop a method that reduces the communication requirements in MG.

5. Try to develop a communication network for the MG in which the communication is performed through satellite medium.

REFERENCES

1. Kumar, Abhishek, Arvind R. Singh, Yan Deng, Xiangning He, Praveen Kumar, and Ramesh C. Bansal. "Multiyear load growth based techno-financial evaluation of a microgrid for an academic institution." *IEEE Access* 6 (2018): 37533–37555.
2. DuVivier, Katharine K. "Mobilizing microgrids for energy justice." *Stanford Technology Law Review* 26 (2022): 250.
3. Kumar, Abhishek, Bikash Sah, Arvind R. Singh, Yan Deng, Xiangning He, Praveen Kumar, and Ramesh C. Bansal. "A review of multi criteria decision making (MCDM) towards sustainable renewable energy development." *Renewable and Sustainable Energy Reviews* 69 (2017): 596–609.
4. Singh, Ranjay, and Ramesh C. Bansal. "Review of HRESs based on storage options, system architecture and optimisation criteria and methodologies." *IET Renewable Power Generation* 12, no. 7 (2018): 747–760.
5. Wiser, Ryan, Mark Bolinger, Ben Hoen, Dev Millstein, Joseph Rand, Galen Barbose, Naïm Darghouth et al. "Land-Based wind Market Report: 2023 Edition." (2023).
6. Xu, Jianyu, Bin Liu, Huadong Mo, and Daoyi Dong. "Bayesian adversarial multi-node bandit for optimal smart grid protection against cyber attacks." *Automatica* 128 (2021): 109551.
7. San, Guocheng, Wenlin Zhang, Xiaoqiang Guo, Changchun Hua, Huanhai Xin, and Frede Blaabjerg. "Large-disturbance stability for power-converter-dominated microgrid: A review." *Renewable and Sustainable Energy Reviews* 127 (2020): 109859.
8. Farrokhabadi, Mostafa, Claudio A. Cañizares, John W. Simpson-Porco, Ehsan Nasr, Lingling Fan, Patricio A. Mendoza-Araya, Reinaldo Tonkoski et al. "Microgrid stability definitions, analysis, and examples." *IEEE Transactions on Power Systems* 35, no. 1 (2019): 13–29.
9. Rathor, Sumit K., and Dipti Saxena. "Energy management system for smart grid: An overview and key issues." *International Journal of Energy Research* 44, no. 6 (2020): 4067–4109.
10. Akinyele, Daniel, Juri Belikov, and Yoash Levron. "Challenges of microgrids in remote communities: A STEEP model application." *Energies* 11, no. 2 (2018): 432.
11. Roslan, Muhammad Farhan., Mahammad Abdul Hannan, Pin Jern Ker, Musfika Mannan, Kashem M. Muttaqi, and Teuku Meurah Indra Mahlia. "Microgrid control methods toward achieving sustainable energy management: A bibliometric analysis for future directions." *Journal of Cleaner Production* 348 (2022): 131340.
12. Ma, Shuyang, Yan Li, Liang Du, Jianzhong Wu, Yue Zhou, Yichen Zhang, and Tao Xu. "Programmable intrusion detection for distributed energy resources in cyber–physical networked microgrids." *Applied Energy* 306 (2022): 118056.
13. Molderink, Albert, Vincent Bakker, Maurice G. C. Bosman, Johann L. Hurink, and Gerard J. M. Smit. "Management and control of domestic smart grid technology." *IEEE Transactions on Smart Grid* 1, no. 2 (2010): 109–119.

14. Vasirani, Matteo, and Sascha Ossowski. "Smart consumer load balancing: state of the art and an empirical evaluation in the Spanish electricity market." *Artificial Intelligence Review* 39 (2013): 81–95.
15. Rajesh, Kurup Sathy, Subhransu Sekhar Dash, Ragam Rajagopal, and Ramasamy Sridhar. "A review on control of ac microgrid." *Renewable and Sustainable Energy Reviews* 71 (2017): 814–819.
16. Guo, Chenyu, Xin Wang, Yihui Zheng, and Feng Zhang. "Real-time optimal energy management of microgrid with uncertainties based on deep reinforcement learning." *Energy* 238 (2022): 121873.
17. Cagnano, A. E. D. T., Enrico De Tuglie, and Pierluigi Mancarella. "Microgrids: Overview and guidelines for practical implementations and operation." *Applied Energy* 258 (2020): 114039.
18. Aristizábal, Andrés Julián, Jorge Herrera, Mónica Castañeda, Sebastián Zapata, Daniel Ospina, and Edison Banguero. "A new methodology to model and simulate microgrids operating in low latitude countries." *Energy Procedia* 157 (2019): 825–836.
19. Sharma, Arvind, Mohan Kolhe, Ulltveit-Moe Nils, Ashwini Mudgal, Kapil Muddineni, and Shirish Garud. "Comparative analysis of different types of micro-grid architectures and controls." In *2018 International Conference on Advances in Computing, Communication Control and Networking (ICACCCN)*, pp. 1200–1208. IEEE, Greater Noida, India, 2018.
20. Vineetha, C. P., and Chembathu Ayyappan Babu. "Smart grid challenges, issues and solutions." In *2014 International Conference on Intelligent Green Building and Smart Grid (IGBSG)*, pp. 1–4. IEEE, Taipei, Taiwan, 2014.
21. Faazila, Fathima S., and L. Premalatha. "Protection strategies for ac and dc microgrid–A review of protection methods adopted in recent decade." *IETE Journal of Research* 69, no. 9 (2023): 6573–6589.
22. Jayamaha, Don Kasun Joseph Shan, N. Widanagama Arachchige Lidula, and Athula D. Rajapakse. "Protection and grounding methods in DC microgrids: Comprehensive review and analysis." *Renewable and Sustainable Energy Reviews* 120 (2020): 109631.
23. Alasali, Feras, Saad M. Saad, Abdelaziz Salah Saidi, Awni Itradat, William Holderbaum, Naser El-Naily, and Fatima F. Elkuwafi. "Powering up microgrids: A comprehensive review of innovative and intelligent protection approaches for enhanced reliability." *Energy Reports* 10 (2023): 1899–1924.
24. Uddin, Moslem, Huadong Mo, Daoyi Dong, Sondoss Elsawah, Jianguo Zhu, and Josep M. Guerrero. "Microgrids: A review, outstanding issues and future trends." *Energy Strategy Reviews* 49 (2023): 101127.
25. Burr, Michael T., Michael J. Zimmer, Brian Meloy, James Bertrand, Walter Levesque, Guy Warner, and John D. McDonald. "Minnesota microgrids." *Minnesota Dept. Commerce* (2013).
26. Hirsch, Adam, Yael Parag, and Josep Guerrero. "Microgrids: A review of technologies, key drivers, and outstanding issues." *Renewable and Sustainable Energy Reviews* 90 (2018): 402–411.

27. Shahzad, Sulman, Muhammad Abbas Abbasi, Hassan Ali, Muhammad Iqbal, Rania Munir, and Heybet Kilic. "Possibilities, challenges, and future opportunities of microgrids: A review." *Sustainability* 15, no. 8 (2023): 6366.
28. Dwivedi, Sanjeet Kumar, Shailendra Jain, Krishna Kumar Gupta, and Pradyumn Chaturvedi. *Modeling and Control of Power Electronics Converter System for Power Quality Improvements.* New York: Academic Press, 2018.
29. Gaur, Prerna, and Sunita Singh. "Investigations on issues in microgrids." *Journal of Clean Energy Technologies* 5, no. 1 (2017): 47–51.
30. Elmetwaly, Ahmed Hussain, Azza Ahmed Eldesouky, and Abdelhay Ahmed Sallam. "An adaptive D-FACTS for power quality enhancement in an isolated microgrid." *IEEE Access* 8 (2020): 57923–57942.
31. Paredes, Luis A., Marcelo G. Molina, and Benjamín R. Serrano. "Resilient microgrids with FACTS technology." In *2020 IEEE PES Transmission & Distribution Conference and Exhibition-Latin America (T&D LA)*, pp. 1–6. IEEE, Montevideo, Uruguay, 2020.
32. Gopakumar, Pathirikkat, M. Jaya Bharata Reddy, and Dusmanta Kumar Mohanta. "Stability concerns in smart grid with emerging renewable energy technologies." *Electric Power Components and Systems* 42, no. 3–4 (2014): 418–425.
33. Karimi, Mohammad Hossein, Seyed Abbas Taher, Zahra Dehghani Arani, and Josep M. Guerrero. "Imbalance power sharing improvement in autonomous microgrids consisting of grid-feeding and grid-supporting inverters." In *7th Iran Wind Energy Conference (IWEC2021)*, pp. 1–6. IEEE, Shahrood, Iran, 2021.
34. Li, Yun Wei, D. Mahinda Vilathgamuwa, and Poh Chiang Loh. "A grid-interfacing power quality compensator for three-phase three-wire microgrid applications." *IEEE Transactions on Power Electronics* 21, no. 4 (2006): 1021–1031.
35. Liang, Xiaodong. "Emerging power quality challenges due to integration of renewable energy sources." *IEEE Transactions on Industry Applications* 53 (2016): 855–866.
36. Awerbuch, Shimon, and Alistair Preston, eds. *The Virtual Utility: Accounting, Technology & Competitive Aspects of the Emerging Industry.* Vol. 26. Cham, Switzerland: Springer Science & Business Media, 2012.
37. Kasaei, Mohammad Javad, Majid Gandomkar, and Javad Nikoukar. "Optimal management of renewable energy sources by virtual power plant." *Renewable Energy* 114 (2017): 1180–1188.
38. Yu, Songyuan, Fang Fang, Yajuan Liu, and Jizhen Liu. "Uncertainties of virtual power plant: Problems and countermeasures." *Applied Energy* 239 (2019): 454–470.
39. Pudjianto, Danny, Charlotte Ramsay, and Goran Strbac. "Virtual power plant and system integration of distributed energy resources." *IET Renewable Power Generation* 1, no. 1 (2007): 10–16.
40. Pau, Marco, Edoardo Patti, Luca Barbierato, Abouzar Estebsari, Enrico Pons, Ferdinanda Ponci, and Antonello Monti. "A cloud-based smart metering infrastructure for distribution grid services and automation." *Sustainable Energy, Grids and Networks* 15 (2018): 14–25.

41. Borghetti, Alberto, Mauro Bosetti, Samuele Grillo, Stefano Massucco, Carlo Alberto Nucci, Mario Paolone, and Federico Silvestro. "Short-term scheduling and control of active distribution systems with high penetration of renewable resources." *IEEE Systems Journal* 4, no. 3 (2010): 313–322.
42. Ehsan, Ali, and Qiang Yang. "State-of-the-art techniques for modelling of uncertainties in active distribution network planning: A review." *Applied Energy* 239 (2019): 1509–1523.
43. Balakumar, P., and Shanmugavel Sathiya. "Demand side management in smart grid using load shifting technique." In *2017 IEEE International Conference on Electrical, Instrumentation and Communication Engineering (ICEICE)*, pp. 1–6. IEEE, Karur, India, 2017.
44. Liu, Weijia, Qiuwei Wu, Fushuan Wen, and Jacob Østergaard. "Day-ahead congestion management in distribution systems through household demand response and distribution congestion prices." *IEEE Transactions on Smart Grid* 5, no. 6 (2014): 2739–2747.
45. Leadbetter, Jason, and Lukas Swan. "Battery storage system for residential electricity peak demand shaving." *Energy and Buildings* 55 (2012): 685–692.
46. Bereczki, Bence, Bálint Hartmann, and Sándor Kertész. "Industrial application of battery energy storage systems: Peak shaving." In *2019 7th International Youth Conference on Energy (IYCE)*, pp. 1–5. IEEE, Bled, Slovenia, 2019.
47. Telaretti, Enrico, and Luigi Dusonchet. "Battery storage systems for peak load shaving applications: Part 1: Operating strategy and modification of the power diagram." In *2016 IEEE 16th International Conference on Environment and Electrical Engineering (EEEIC)*, pp. 1–6. IEEE, Florence, Italy, 2016.
48. Silva, Bhagya Nathali, Murad Khan, and Kijun Han. "Futuristic sustainable energy management in smart environments: A review of peak load shaving and demand response strategies, challenges, and opportunities." *Sustainability* 12, no. 14 (2020): 5561.
49. Gharavi, Hamid, and Bin Hu. "Multigate communication network for smart grid." *Proceedings of the IEEE* 99, no. 6 (2011): 1028–1045.
50. Barmada, Sami, Antonino Musolino, Marco Raugi, Rocco Rizzo, and Mauro Tucci. "A wavelet based method for the analysis of impulsive noise due to switch commutations in power line communication (PLC) systems." *IEEE Transactions on Smart Grid* 2, no. 1 (2011): 92–101.
51. Butt, Osama Majeed, Muhammad Zulqarnain, and Tallal Majeed Butt. "Recent advancement in smart grid technology: Future prospects in the electrical power network." *Ain Shams Engineering Journal* 12, no. 1 (2021): 687–695.
52. Zhou, Quan, Mohammad Shahidehpour, Aleksi Paaso, Shay Bahramirad, Ahmed Alabdulwahab, and Abdullah Abusorrah. "Distributed control and communication strategies in networked microgrids." *IEEE Communications Surveys & Tutorials* 22, no. 4 (2020): 2586–2633.
53. Katiraei, Farid, Reza Iravani, Nikos Hatziargyriou, and Aris Dimeas. "Microgrids management." *IEEE Power and Energy Magazine* 6, no. 3 (2008): 54–65.

54. Zhang, Jianhua, Adarsh Hasandka, S. M. Shafiul Alam, Tarek Elgindy, Anthony R. Florita, and Bri-Mathias Hodge. "Analysis of hybrid smart grid communication network designs for distributed energy resources coordination." In *2019 IEEE Power & Energy Society Innovative Smart Grid Technologies Conference (ISGT)*, pp. 1–5. IEEE, Washington, DC, USA, 2019.

55. Aalamifar, Fariba, Lutz Lampe, Sara Bavarian, and Eugene Crozier. "WiMAX technology in smart distribution networks: Architecture, modeling, and applications." In *2014 IEEE PES T&D Conference and Exposition*, pp. 1–5. IEEE, Chicago, IL, USA, 2014.

56. Avendaño, Jorge Luis Sosa, and Luz Stella Moreno Martín. "Communication in microgrids." *Microgrids Design and Implementation* 49 (2019): 69–96.

57. Lo, Chun-Hao, and Nirwan Ansari. "The progressive smart grid system from both power and communications aspects." *IEEE Communications Surveys & Tutorials* 14, no. 3 (2011): 799–821.

58. Pagani, Pascal, and Andreas Schwager. "A statistical model of the in-home MIMO PLC channel based on European field measurements." *IEEE Journal on Selected Areas in Communications* 34, no. 7 (2016): 2033–2044.

59. Salem, Abdelhamid, Khairi Ashour Hamdi, and Emad Alsusa. "Physical layer security over correlated log-normal cooperative power line communication channels." *IEEE Access* 5 (2017): 13909–13921.

60. N'cho, Janvier Sylvestre, and Issouf Fofana. "Review of fiber optic diagnostic techniques for power transformers." *Energies* 13, no. 7 (2020): 1789.

61. Behnke, Ilja, and Henrik Austad. "Real-time performance of industrial IoT communication technologies: A review." *IEEE Internet of Things Journal* 11 (2023): 7399–7410.

62. IETRE Series. "Microgrids and active distribution networks." *The Institution of Engineering and Technology* 332 (2009): 1568.

63. Shaukat, Neelofar, Md Rabiul Islam, Md Moktadir Rahman, Bilal Khan, Basharat Ullah, Sahibzada Muhammad Ali, and Afef Fekih. "Decentralized, democratized, and decarbonized future electric power distribution grids: A survey on the paradigm shift from the conventional power system to micro grid structures." *IEEE Access* 11 (2023): 60957–60987.

64. Dragičević, Tomislav, Pierluigi Siano, and S. R. Sahaya Prabaharan. "Future generation 5G wireless networks for smart grid: A comprehensive review." *Energies* 12, no. 11 (2019): 2140.

65. Bag, Gargi, Linus Thrybom, and Petri Hovila. "Challenges and opportunities of 5G in power grids." *CIRED-Open Access Proceedings Journal* 2017, no. 1 (2017): 2145–2148.

66. Lefred-Rivera, Erick, Eduardo Marlés-Sáenz, and Eduardo Gómez-Luna. "Requirements and tests of the control system of a microgrid, according to the IEEE Std 2030-7-2017 and IEEE Std 2030-8-2018 standards." *Ingeniería y Competitividad* 24, no. 2 (2022).

67. M Series. "IMT vision–framework and overall objectives of the future development of IMT for 2020 and beyond." *Recommendation ITU* 2083, (2015): 1–21.

68. Varghese, Anitha, Deepaknath Tandur, and Apala Ray. "Suitability of WiFi based communication devices in low power industrial applications." In *2017 IEEE International Conference on Industrial Technology (ICIT)*, pp. 1307–1312. IEEE, Toronto, ON, Canada, 2017.

69. Arbab-Zavar, Babak, Emilio J. Palacios-Garcia, Juan C. Vasquez, and Josep M. Guerrero. "Smart inverters for microgrid applications: A review." *Energies* 12, no. 5 (2019): 840.

70. Zhou, Weiwei, Ziyuan Tong, Zhao Yang Dong, and Yu Wang. "LoRa-Hybrid: A LoRaWAN Based multihop solution for regional microgrid." In *2019 IEEE 4th International Conference on Computer and Communication Systems (ICCCS)*, pp. 650–654. IEEE, Singapore, 2019.

71. Wen, Miles H. F., Ka-Cheong Leung, Victor O. K. Li, Xingze He, and Chung-Chieh Jay Kuo. "A survey on smart grid communication system." *APSIPA Transactions on Signal and Information Processing* 4 (2015): e5.

72. IEEE Standards Association. "2030-2011 IEEE guide for smart grid interoperability of energy technology and information technology operation with the electric power system (EPS), and end-use applications and loads." *IEEE Smart Grid. Retrieved*, 2013.

73. Photovoltaics, Dispersed Generation, and Energy Storage. "*IEEE Guide for Monitoring, Information Exchange, and Control of Distributed Resources Interconnected with Electric Power Systems.*" New York: IEEE, 2007.

74. International Electrotechnical Commission. "*IEC 62351 Power Systems Management and Associated Information Exchange—Data and Communications Security.*" Geneva, Switzerland: International Electrotechnical Commission, 2007.

75. Brunner, Christoph. "IEC 61850 for power system communication." In *2008 IEEE/PES Transmission and Distribution Conference and Exposition*, pp. 1–6. IEEE, Chicago, IL, USA, 2008.

76. Sánchez, Gemma, Isabel Gómez, Joaquín Luque, Jaime Benjumea, and Octavio Rivera. "Using internet protocols to implement IEC 60870-5 telecontrol functions." *IEEE Transactions on Power Delivery* 25, no. 1 (2009): 407–416.

77. International Electrotechnical Commission. "IEC 61968-1 Application integration at electric utilities–system interfaces for distribution management Part 1: Interface architecture and general requirements." *IEC Reference number IEC* (2003): 61968-1.

78. Zhong, Xingsi, Lu Yu, Richard Brooks, and Ganesh Kumar Venayagamoorthy. "Cyber security in smart DC microgrid operations." In *2015 IEEE First International Conference on DC Microgrids (ICDCM)*, pp. 86–91. IEEE, Atlanta, GA, USA, 2015.

79. Wang, Wenye, and Zhuo Lu. "Cyber security in the smart grid: Survey and challenges." *Computer Networks* 57, no. 5 (2013): 1344–1371.

CHAPTER 3

Hierarchical Control of Microgrid

3.1 HIERARCHICAL CONTROL STRUCTURE

MG offers several benefits, such as increasing power system reliability, reducing peak demand, reducing GHG emissions, having a prosumer-friendly architecture, integrating RESs, having a scalable nature, reducing energy losses, providing ancillary services, and improving power quality. Due to these advantages, an MG has gained significant interest in the past decade. However, compared to conventional power systems, the nature of generation in MG is intermittent and has a low generated voltage and inertia. As a result, numerous issues arise, such as frequency deviation, system instability, voltage fluctuations, and supply-demand imbalances [1]. Due to these uncertainties, an MG is operated under highly stressed conditions; thus, a robust and efficient control strategy is required to guarantee an efficient and reliable operation [2].

MG has the ability to operate in IS mode as well as in GC mode; therefore, proper control of MG is very important for efficient and stable operation. There are some basic requirements of the MG control that must be fulfilled by the controller; these are (a) DG coordination, (b) regulation of frequency and voltage in both operating modes, (c) reconnection of IS MG with the UG, (d) accurate load sharing among the DGs, (e) power flow between MG and UG, (f) optimization of the operating cost of MG, (g) ensuring system stability during mode switching, and (h) handling and control of transients [3,4].

DOI: 10.1201/9781003594284-3

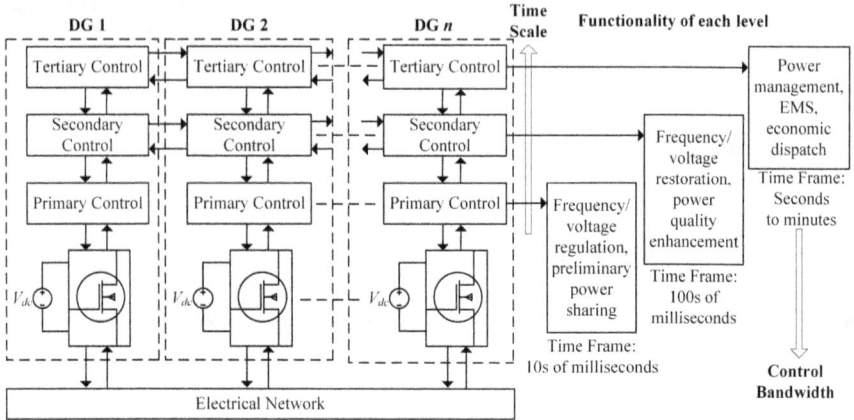

FIGURE 3.1   Schematic of MG hierarchical control structure.

These control requirements have different time scales and are of different significance; thus, to address these requirements at different levels, a hierarchal control structure is needed. A hierarchical structure is used to control an MG that consists of three levels, i.e., primary, secondary, and tertiary control levels. A primary control is the basic level having the minimum decision time step and is concerned with the MG stability and power-sharing within the network. The main objective of SC is to eliminate the voltage and frequency deviations caused by the primary controller. Moreover, recently, more control objectives such as voltage unbalance, harmonic compensation, and reactive power sharing have been introduced in the additional SC loops [1]. The final level of the hierarchy is the tertiary control level, which is concerned with global economic optimization depending on energy prices and current markets [2]. A detailed description of these three control levels is presented in the following sections. The schematic of the hierarchical control structure, along with the different functionality of every level and the timeframe to complete the tasks, are presented in Figure 3.1.

3.2 PRIMARY CONTROL

Among all the control levels in the hierarchy, this level has the fastest response time and sends the control signals at intervals of milliseconds to the inverter to perform the control actions. Despite not being completely standardized, generally, a primary control level consists of inverter output and power-sharing controllers. The inverter output control consists of inner and outer control loops that regulate the current and voltage. A power-sharing controller regulates the frequency and voltage while ensuring accurate power sharing [5]. A generalized schematic of this control level is presented in Figure 3.2. In this section, different inverter output and power-sharing controllers used in an MG will be discussed in detail and are presented in Figure 3.3.

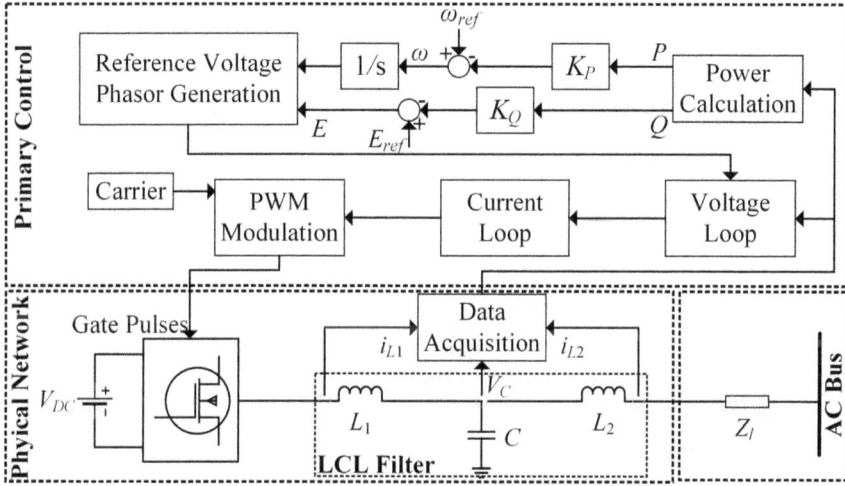

FIGURE 3.2   Block diagram of primary control method.

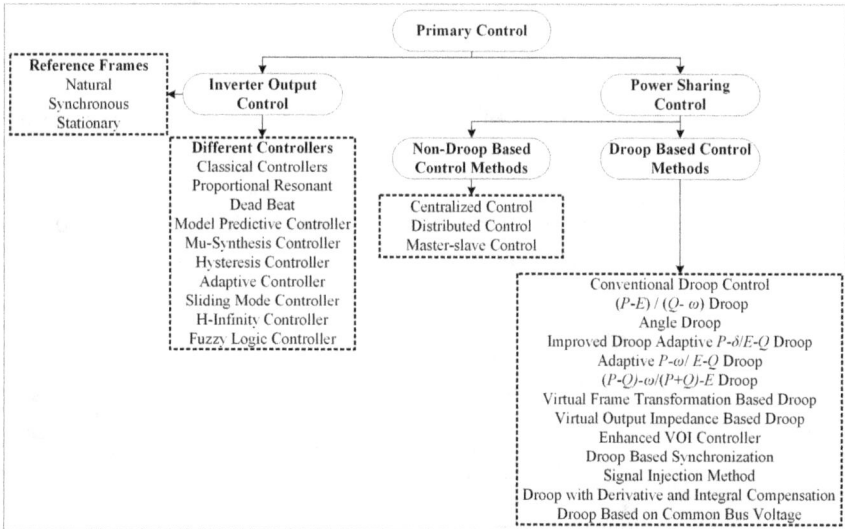

FIGURE 3.3   Classification of primary controller.

3.2.1 Inverter Output Control

In some literature, this control level is also referred to as the zero control level. In this, the inner and outer control loops cascaded one another in a decoupled manner to independently control the voltage and current of the inverter. Therefore, the dynamic speed of the current loop is usually 5–20 times faster than the voltage loop for stability purposes [6]. Moreover, these control loops must not be necessary to be two in number. They can vary according to the control variables (i.e. a single loop will be used in case of one controlling variable and two loops will be used for two controlling variables). This control level is generally implemented using three different reference frames, i.e., natural (abc), synchronous (dq), and stationary ($\alpha\beta$).

3.2.1.1 Reference Frames

This control level is generally implemented using three reference frames. These are:

- **abc Reference Frame:** This reference frame is applied without any transformation to a 3-Φ system; however, a separate controller will be used for every grid current. Moreover, the star or delta connections of the system must be considered in designing process [7]. A generalized schematic two-loop control system employed in abc frame is presented in Figure 3.4 [6]. From Figure 3.4, it can be seen that in the outer voltage control loop, a PI controller is used that generates a reference current (d-component i.e. $i_{d\,ref}$). This d-component of current is then transformed into abc reference frame to get $i_{a\,ref}$, $i_{b\,ref}$, and $i_{c\,ref}$. These reference currents are then compared with the output current of the inverter and are then sent to the inner current control loop. In the inner loop, different controllers, such as hysteresis, PI, and deadbeat, can be used to generate the switching pattern for the inverter.

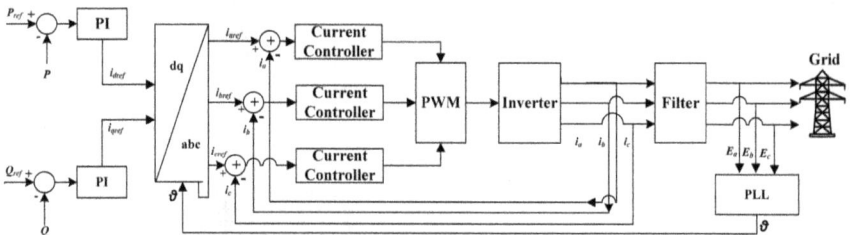

FIGURE 3.4 abc reference frame configuration.

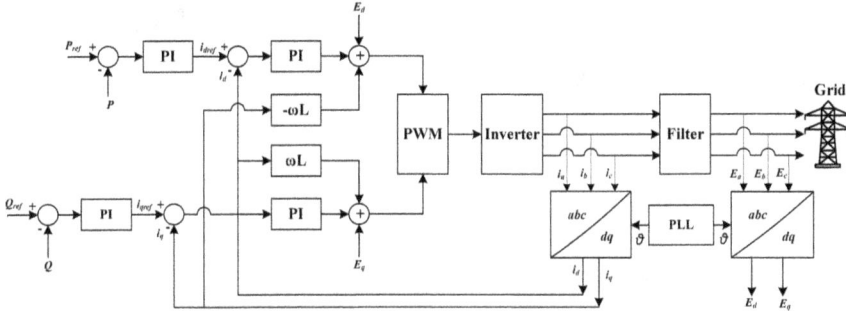

FIGURE 3.5 *dq* reference frame configuration.

- *dq* **Reference Frame:** In *dq*-reference frame, the voltage and current from *abc* frame are converted *dq* frame using Park transformation. Thus, the sinusoidal current and voltage are transformed into DC domain (current (i_d and i_q) and voltage (E_d and E_q) which can be regulated and filtered easily [8]. A generalized schematic single-loop control structure employed in *dq* frame is presented in Figure 3.5 [6,9]. The current components (i_d and i_q) are compared with the reference current (i_{dref} and i_{qref}) components. From the comparison an error is generated that is then fed to a PI controller. The output of the controllers is added with the voltage components (E_d and E_q) and $\pm\omega L$; and are then fed to the PWM modulator to generate the signals that drive the inverter switches.

- *αβ* **Reference Frame:** The voltages (E_a, E_b, and E_c) and current (i_a, i_b, and i_c) from *abc* frame are transformed into voltage (E_α and E_β) and current (i_α and i_β) of *αβ* reference frame using the Clark transformation, respectively. A generalized schematic of the Proportional Resonant (PR) controller applied in a decoupled manner in *αβ* frame is shown in Figure 3.6 [6,10]. Instead of two control loops, the schematic presented in Figure 3.6 has three control loops. A first control loop is used to regulate the current injected into the grid; a second loop is used to regulate the capacitor voltage of the filter; and a third loop is used to regulate the inductor current of the filter. These control variables are initially converted into sinusoidal variables due to high tracking capability of the Proportional Resonant (PR) controller to track the sinusoidal references. The output of the PR controller is then fed to the PWM module to generate the desired output for the switches of the inverter.

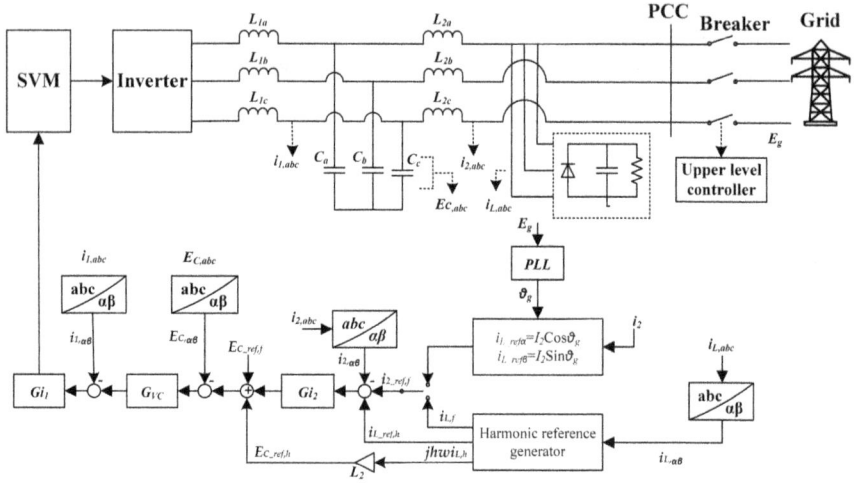

FIGURE 3.6 $\alpha\beta$ reference frame configuration.

3.2.1.2 Controllers Used in Inverter Output Control

In a state-of-the-art literature review, different controllers are used in these reference frames to provide high-quality output waveforms. Some of the controllers discussed in the literature are described below:

- **Classical Controllers:** Classical controllers such as PI, PID, P, and PD controllers have been most commonly used in different applications over the last few years due to their simple structure, realization, and implementation. Moreover, these controllers are applicable in single input single output systems, linear time-invariant systems, etc. [11]. These controllers not only respond to the DC variables but sometimes also track the sinusoidal references, as the authors in Ref. [12] use a PI current controller with integrators associated with $\alpha\beta$ -frame and applied it under distorted and undesirable operating conditions. It results in zero steady-state error with a low computation burden concerning current harmonics.

- **PR Controller:** Compared to the classical controllers that have low capability to track the sinusoidal references, a PR controller has high tracking capability to track the sinusoidal references. Therefore, PR controllers are more likely to be employed in $\alpha\beta$ and *abc* reference frames [13]. Moreover, a second major advantage of PR controllers over classical controllers is that in a PR controllers only the frequencies that are close to the resonant frequencies are integrated; thus, as a result, no stationary magnitude and phase shift are involved [14]. The authors in Ref. [15] used a PR controller in an inner current control loop in a distributed power system and concluded that it shows efficient performance and results in high gain around the resonant frequency. Moreover, it shows fast dynamic performance with zero steady-state error and no excessive computational burden.

- **Dead Beat Controller:** It belongs to a family of predictive controllers and is widely used in different modern applications due to its fast dynamic response. This controller has the ability to place the close loop poles at zero; thus, within a few sampling steps, the tracking error becomes zero that makes it very feasible for current regulation at the zero control level [16]. The authors in Ref. [17] used a PI controller in the outer voltage loop and a dead beat controller in the inner current loop to regulate the voltage and current of the system respectively and provide high-quality output waveforms. This controller shows high tracking capability and can easily handle parameter mismatching or system disturbances.

- **Model Predictive Controller (MPC):** An MPC is a sophisticated control process that uses a system model to predict the behaviour of the controlled variables in future. Moreover, it has a high capability to efficiently handle system non-linearities and track the reference accurately. Numerous authors have designed different MPC for different power system applications; such as the authors in Ref. [18] have designed a finite set-based MPC for multi-level inverter-based GC MG. The controller tracks its reference efficiently with fast dynamic response, and high-quality output waveforms are attained. Similarly, an MPC for two-level inverter-based MG to regulate the output current and voltage is proposed in Ref. [19]. Although, MPCs show very good performance and can handle system uncertainties efficiently, they are difficult to implement and require huge calculations [20].

- **Mu-Synthesis Controller:** This controller is based on a control theory related to system uncertainties. The concept of this controller design is based on a singular value that can either be unstructured or structured. This controller shows efficient performance in case of system error and ensures system stability in single as well as multi-variable systems. This controller can be used in numerous power system applications such as the authors in Ref. [21] used it in a 3-Φ inverter-based MG system. Similarly, the authors in Ref. [22]

proposed a mu-synthesis that is applied to inverter-based MG in *dq* reference frame. This controller has the capability to handle the unknown and time-varying load and shows flexible performance in case of system uncertainties.

- **Hysteresis Controller:** In this approach, the controller is attached to an adaptive band in order to attain a fixed switching frequency. The output is the switches' state that provides variable switching frequency. This controller has numerous advantages, such as simplicity, fast dynamic response, load independence, and robustness to varying parameters [23]. However, a main limitation of this approach is that it limits the current within its band limits, which causes unwanted variation in switching frequency. To overcome this limitation, numerous algorithms and techniques are present in the literature, a detailed explanation of which can be found in Ref. [24]. The authors in Ref. [25] designed a hybrid controller that combines a hysteresis controller with Sliding Mode Controller (SMC) for inverter-based GC MG. In this technique, the authors used a single-order SMC that enables the controller to attain a stable switching frequency. This controller generates the switching pulses for the inverter by comparing the instantaneous grid current with the reference current. However, it is not able to fully control the grid-injected current and the system faces the chattering effect.

- **Adaptive Controllers:** Depending on the operating conditions of the system, this controller has the capability to automatically adjust its control actions. This controller is widely used in numerous MG applications, such as the authors in Ref. [26] combined an adaptive controller with the deadbeat controller to regulate the output current of the inverter. In this approach, both controllers perform their control actions in parallel, thus regulating the output current with a fast dynamic response. Similarly, the authors in Ref. [27] designed a non-linear adaptive controller to control the active and reactive power flow in GC inverter. A controller shows significant performance in different atmospheric conditions.

- **SMC:** This controller shows high robustness against variable variation and load disturbances; therefore, it is widely used in power system applications. In case of any deviation from its reference, the controller takes significant control actions and achieves its reference point with fast dynamic response. However, beside these numerous advantages, there are some drawbacks of SMC, such as: (a) The selection of a suitable sliding surface that enables the SMC to perform its control actions is a difficult task, (b) the performance of the controller is greatly degraded if the sampling time is not chosen carefully, and (c) during the tracking process, the SMC results in high chattering [28]. Hence, to improve the quality of output waveforms and remove the chattering effect, the authors in Ref. [29] combined SMC with Fuzzy Logic Controller (FLC). In this approach, an SMC is used to control the inverter performance while FLC is used to approximate the unwanted disturbances that happen due to variation in the atmosphere.

- **H∞ Controller:** This controller is feasible to solve multi-variable problems; however, its solution is only possible when the problem is presented as an optimized problem. It has the capability to inject pure sinusoidal current to the UG in case of linear and non-linear loads. To regulate the current, the authors in Ref. [30] proposed a hybrid approach in which H∞ is combined with a repetitive controller. This hybrid approach shows efficient steady-state and dynamic response with and without loads. Moreover, it shows high performance in suppressing total harmonic distortion, however, this approach shows slow dynamic response and has computational complexity. Similarly, the authors in Ref. [31] proposed an H∞-based control strategy to improve the quality of grid-injected current and optimal current flow in both IS and GC modes. High-quality current waveforms are achieved even with non-linear and unbalanced loads.

- **FLC:** In FLC, the control parameters are defined by using the knowledge of the human being and are then used to control the dynamics of the system. There are four main components of FLC approach; i.e., (a) rule base, (b) fuzzification, (c) interference mechanism, and (d) de-fuzzification. FLCs are widely used in numerous power system applications such as the authors in Ref. [32] used FLC in the current control loop of the inverter to regulate the current. This approach is implemented in a dq reference frame and tracks the current very precisely with no overshoot. Similarly, the authors in Ref. [33] used a combination of FLC and a back-stepping controller to regulate the DC-link voltage in a 3-Φ GC PV inverter. This controller efficiently regulates the DC-link voltage, and with a unity power factor, it delivers power to the UG.

3.2.1.3 Example of Implementation of Inverter Output Control

Consider a two stage PV configuration system in which a PV MPPT technique and DC-DC converter are employed in a first stage while DC-AC inverter is employed in a second stage as shown in Figure 3.7.

In this work, the same MPPT technique and DC-DC converter are used in a first stage which are discussed in Example 1–6. However, more PV panels are connected in series and parallel combination to make it feasible for the inverter-based DG application. Moreover, in a second stage a universal inverter is used that not only perform conversion but also regulate voltage and current to attain the high-quality output waveforms. The control of the inverter is usually implemented in cascaded manner that consists the voltage and current control loops. In this case scenario, a voltage loop is used to regulate a constant voltage at the DC-link while a current loop is responsible for the injection of high-quality current into the UG as shown in Figure 3.8.

From Figure 3.8, it can be seen that initially, the measured and reference values of the DC-link voltages are compared to generate an error. The error

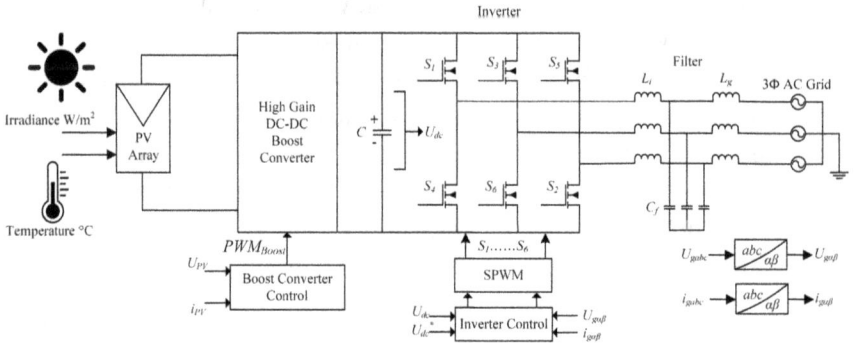

FIGURE 3.7   Schematic of a 2-stage 3Φ grid-connected PV system.

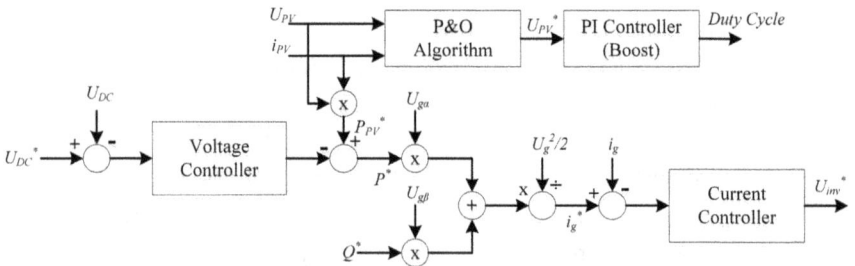

FIGURE 3.8   Generalized schematic of control structure.

is fed to the voltage control loop where a PI, adaptive PI, fractional order PI, and adaptive fractional order PI controller is used for DC-link voltage regulation [34]. Moreover, a PR controller is implemented in the inner current control loop that generate the voltage reference. The desired voltage generated by the current controller is then fed to phase disposition-based PWM technique (discussed in Example 1–7) that generate the desired signals for the switches of the inverter.

To validate the performance of the whole PV system (in prospect of this book, it can be called a DG) during rapidly varying irradiance and temperature scenario simulations are performed in MATLAB® environment. The irradiance and temperature varies at every 1 sec, with these variations, the output voltage of the PV panel also changes, as shown in Figure 3.9a and b. A DC-DC converter step-up the low generated voltage to a high level, and a DC-link voltage controller is used to regulate the voltage at 700 V. When the irradiance and temperature level changes, a fluctuation in DC-link voltage can be observed, but they are restored by the controllers effectively as presented in Figure 3.9c. Similarly, a 3Φ grid injected current during varying environmental conditions is shown in Figure 3.9d while a single phase grid injected current is presented in Figure 3.9e. From Figure 3.9e, it can be observed that high-quality waveforms are achieved at the output of inverter with very low total harmonic distortion.

3.2.2 Power-Sharing Control

This controller is responsible for regulating the frequency and voltage while ensuring appropriate power-sharing [35]. In practical systems, the DGs are physically distributed and their power-sharing controller is mainly based on local measurements; therefore, their control can be implemented with or without the use of CN [36]. Hence, based on the requirement of CN, it can be categorized into droop and non-droop-based control methods, as

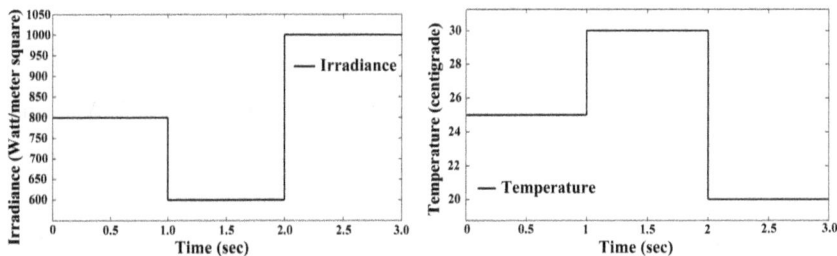

FIGURE 3.9a   Simulation results of PV panel irradiance and temperature.

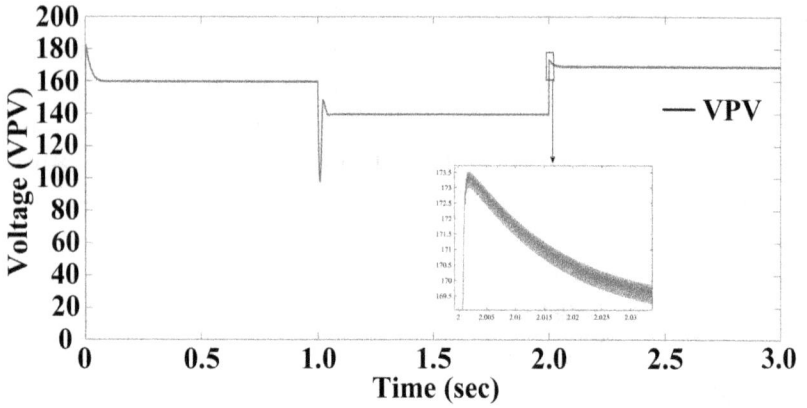

FIGURE 3.9b   Simulation results of PV panel voltage.

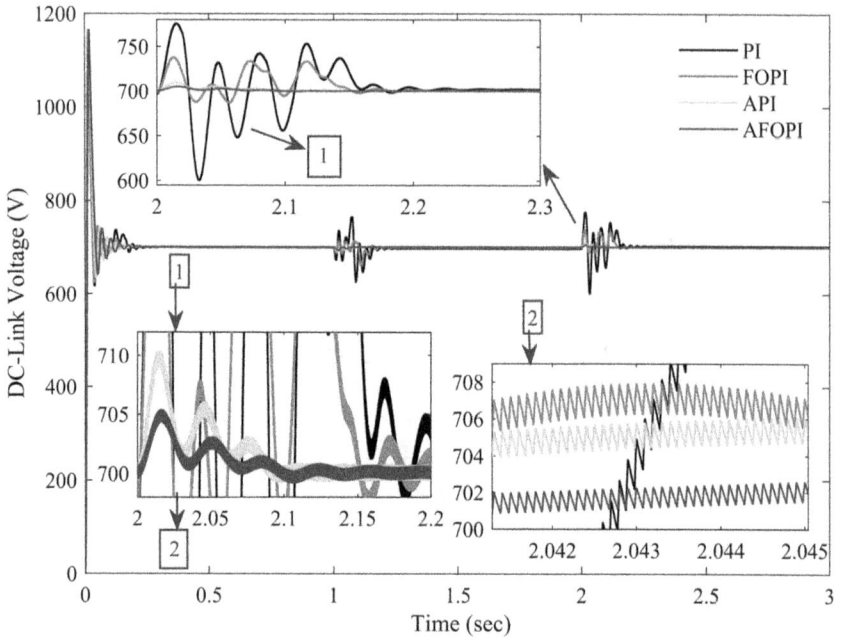

FIGURE 3.9c   Simulation results of DC-link voltage using PI, FOPI, API, and AFOPI.

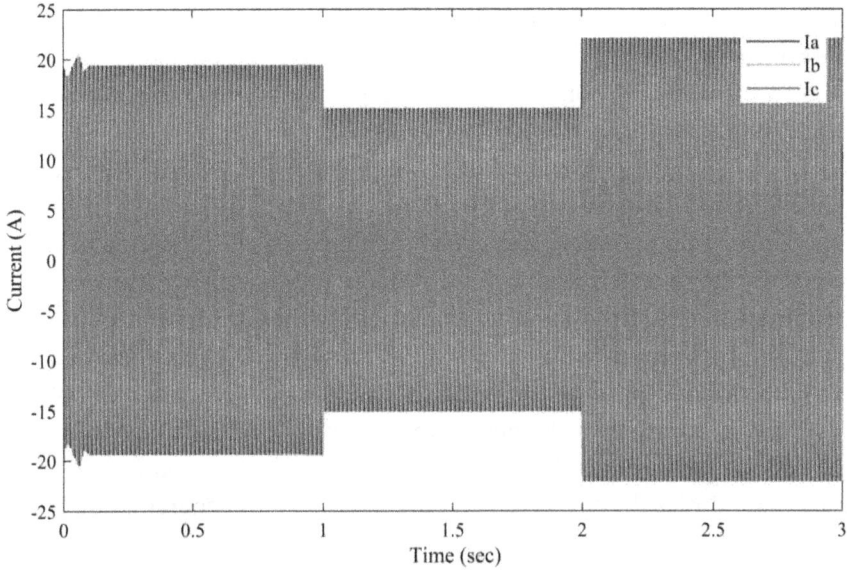

FIGURE 3.9d   Simulation results of 3Φ grid injected current.

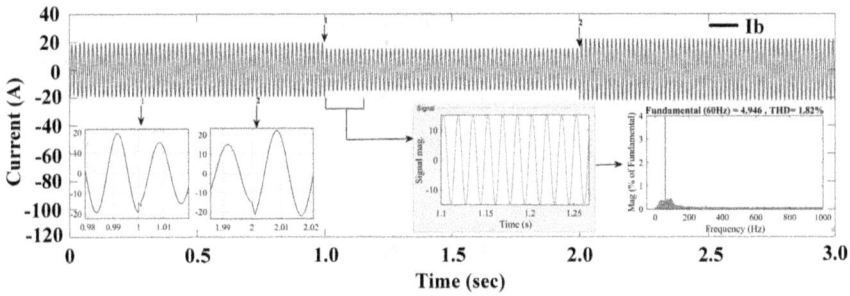

FIGURE 3.9e   Simulation results of 1Φ grid injected current with THD analysis.

presented in Figure 3.3. The droop-based control methods are independent of the inter-unit cyber channel. The droop-based control methods are easy to implement, have high redundancy, and are more flexible, but there are some disadvantages, such as inaccurate power sharing, circulation of current among DGs, and slow dynamic response. The non-droop-based control methods have the advantages of accurate power sharing, good power quality, and fast dynamic response, but they impose high costs, reduce the system expansion flexibility, and high implementation complexity [37]. Different droop and non-droop-based control methods are discussed in detail in this sub-section.

3.2.2.1 Droop-Based Control Methods

In a conventional power system, a synchronous generator is used to regulate the voltage and frequency and provide power-sharing features. The imbalance between the input mechanical power of the synchronous generator and its output active power (due to electromagnetic field) causes a change in the speed of the rotor that appears as frequency deviation. Likewise, a variation in the output reactive power results in variation in the voltage magnitude. Hence, in RES-based DGs, a droop controller is used that mimics the characteristics of a synchronous generator due to the absence or very low inertia.

- **Conventional Droop Control:** The most commonly used droop controllers are Active Power-Frequency (P-ω) droop and Reactive Power-Voltage (Q-E) droop controllers and are discussed in Ref. [38]. These droop controllers are integrated with the Voltage Source Inverters (VSIs) to attain good power sharing and can be expressed as [39]:

$$\omega = \omega_{\text{ref}} - K_P P \tag{3.1}$$

$$E = E_{\text{ref}} - K_Q Q \tag{3.2}$$

where ω and E are the operating angular frequency and measured output voltage of the DG respectively, ω_{ref} and E_{ref} are the nominal angular frequency and voltage respectively, K_P and K_Q are the active and reactive power droop coefficients and can be calculated using (3.3) and (3.4) respectively [40,41].

$$K_P = \frac{\omega_{\max} - \omega_{\min}}{P_{\max}} \tag{3.3}$$

$$K_Q = \frac{E_{\max} - E_{\min}}{Q_{\max}} \tag{3.4}$$

where the subscripts max and min present the maximum and minimum of the respective variables. The characteristic slopes of droop

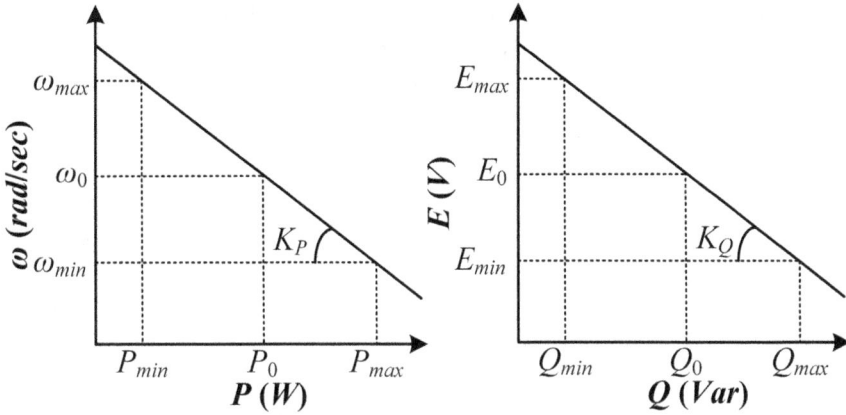

FIGURE 3.10 P-ω and Q-E droop curves.

controllers are presented in Figure 3.10 [42]. Besides, easy and simple implementation, it has some disadvantages, which are listed below:

1. A droop controller is applicable only for the predominant inductive DG interconnecting lines and transmission lines. However, in physical systems, the low and medium feeders are mainly resistive [43,44].

2. Under critical load, it accurately manages the reactive power sharing but results in week voltage regulation. As voltage is not a global variable in an MG network therefore accurate reactive power sharing is a challenge as it results in circulating reactive current [45,46].

3. It is not feasible for non-linear loads, as only the fundamental components of current and voltage are taken into account and only the averaged value of P and Q are measured while neglecting the harmonics. The non-linear loads result in harmonic current circulation that results in poor power quality [47,48].

4. There is a trade-off between accurate power sharing and frequency/voltage deviation. A steeper droop curve results in better power sharing, but at the same time, a large deviation in voltage and frequency from their references is observed [46].

To cope with the above-mentioned drawbacks of the traditional droop controller, numerous modifications have been made to the controller that are discussed below in detail.

- **P-E and Q-ω Droop Control:** A conventional droop controller is suitable for high inductive distribution feeders; however, the feeders are mainly resistive in low distribution networks. Therefore, in such scenarios, the P-ω/Q-E droop controllers did not work efficiently and failed to maintain appropriate power sharing. To cope with this challenge, the authors in Ref. [49] proposed P-E and Q-ω droop controllers for highly resistive impedance lines that can expressed as:

$$E = E_{\text{ref}} - K_P P \tag{3.5}$$

$$\omega = \omega_{\text{ref}} - K_Q Q \tag{3.6}$$

where K_P and K_Q are the droop coefficients of active and reactive respectively. The droop slopes E-P/Q-ω are presented in Figure 3.11. As compared to conventional droop, this method shows high performance in low voltage resistive networks. However, this method has a dependency on MG parameters, as a result, it is unable to accurately resolve the issues such as accurate power-sharing, slow dynamic response, and dependency on line impedance [50].

- **Angle Droop Control:** A conventional droop controller pushes the frequency from its nominal values; therefore, to minimize the deviation, the authors in Ref. [51] presented an angle droop controller. In this controller, the phase angle between the DG's output voltage and the reference frame is used to attain proportionate power sharing among the DGs. Therefore, instead of frequency, the voltage phase angle (δ) is associated with P. The δ can be calculated as:

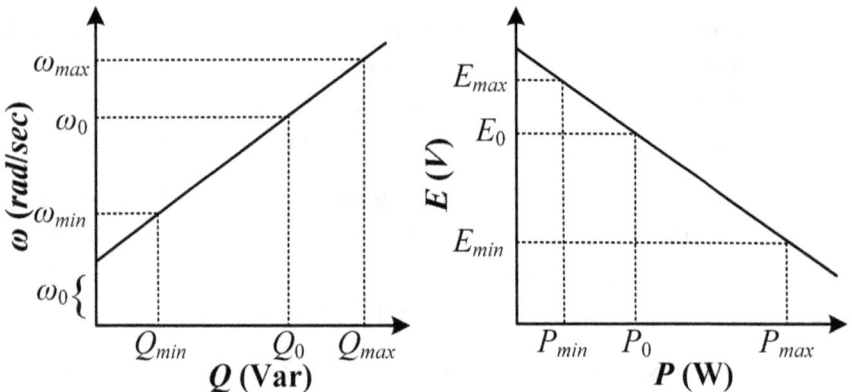

FIGURE 3.11   P-E and Q-ω droop curves.

$$\delta = \int \omega \, dt \tag{3.7}$$

The δ and E equations for P and Q sharing can be presented as:

$$\delta = \delta_{ref} - K_P (P - P_{ref}) \tag{3.8}$$

$$E = E_{ref} - K_Q (Q - Q_{ref}) \tag{3.9}$$

where δ_{ref} is the nominal phase angle, P_{ref} and Q_{ref} are nominal active and reactive power operating at nominal δ_{ref} and E_{ref} respectively. This method shows efficient performance to remove the steady-state frequency deviation as compared to the conventional droop technique; as a result, accurate power sharing is achieved. However, this method may cause instability in the system when the local controllers fail to properly synchronize the power converter with each other [52].

- **Improved Droop Control:** The authors in Ref. [53] described $P - \delta / Q - E$ droop controller, which is similar to the virtual impedance controller. Moreover, it is also explained that the frequency droop control scheme with power derivative feedback is also similar to the virtual impedance controller. Therefore, a large droop coefficient is introduced in this method to ensure improved power-sharing. It means that the effect of the impedance mismatch in a distribution network is mitigated by the large virtual impedance. Hence, by adopting the frequency and virtual impedance droops, an improved controller can be formed and can be given as:

$$\omega_0 = \omega_{ref} - D_P \frac{dP}{dt} \tag{3.10}$$

The virtual impedance and frequency droops are combined to give us a resultant improved droop and can be expressed as:

$$\omega_0 = \omega_{ref} - K_P P - D_P \frac{dP}{dt} \tag{3.11}$$

$$E_0 = E_{ref} - (K_Q + D_Q) Q \tag{3.12}$$

This method improves the dynamic response of the controller and shows more robustness to parameter variations; however, it requires high bandwidth for implementation.

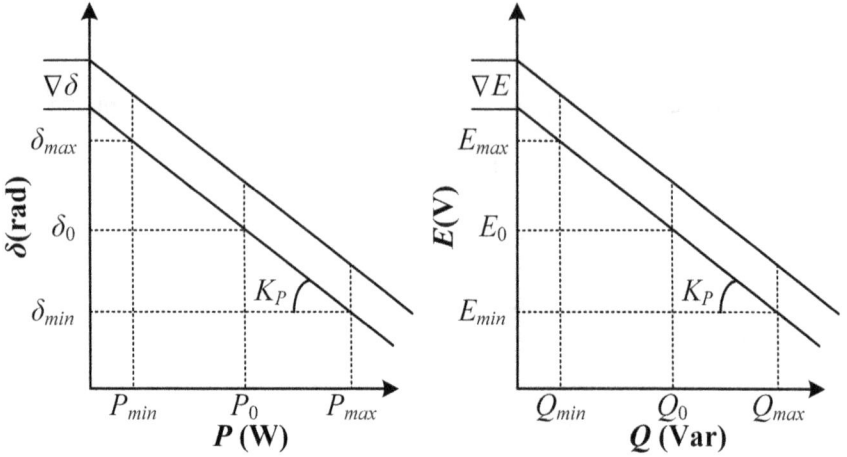

FIGURE 3.12 Adaptive P-δ/E-Q droop curves.

- **Adaptive P-δ/E-Q Droop:** To avoid the high droop gain and its impact on system stability, the authors in Ref. [54] proposed an adaptive $P - \delta / Q - E$ droop controller for power-sharing. In this controller, a supplementary terms, i.e. $\nabla\delta$ in active power sharing and ∇E in reactive power sharing are added to ensure the stability of the close loop control system over a wider range. The droop curves of proposed controller are presented in Figure 3.12 and its expressions can be written as:

$$\delta = \delta_{\text{ref}} - K_P(P - P_{\text{ref}}) + \nabla\delta \tag{3.13}$$

$$E = E_{\text{ref}} - K_Q(Q - Q_{\text{ref}}) + \nabla E \tag{3.14}$$

- **Adaptive P-ω/Adaptive E-Q Droop:** The authors in Ref. [44] proposed an adaptive droop controller for parallel operating DGs. In this method, two additional supplementary terms are added to the equation of the reactive power droop. One term is responsible for compensating the voltage drop that occurs in the feeder, while a second term is responsible for maintaining accurate reactive power sharing. A simplified structure of the two parallel DGs that supply power to the load in an MG is presented in Figure 3.13; and their expression for ith-DER can be written as:

$$V_i = E_{i,\text{ref}} - K_{Qi}Q_i - \frac{\gamma_i P_i}{E_i} - \frac{\chi_i Q_i}{E_i} \tag{3.15}$$

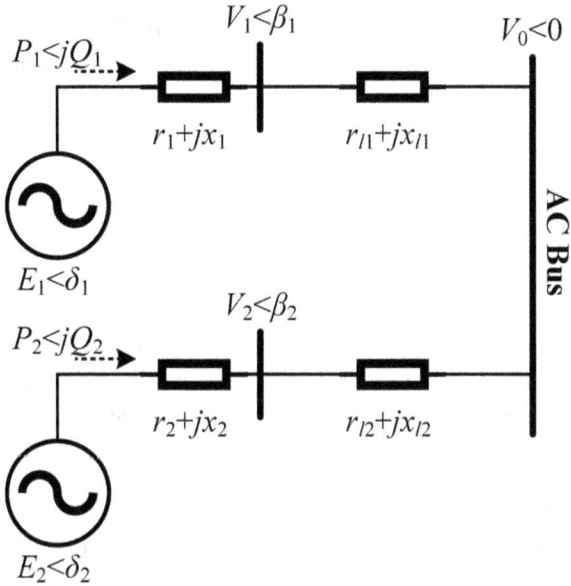

FIGURE 3.13 Two parallel DERs supply power to load.

where γ_i presents the resistance of the feeder while χ_i presents the DER reactance. By putting the last two terms of the above equation in (3.2), we get:

$$E_i = E_{i,\text{ref}}^* - K_{Qi}Q_i + \frac{\gamma_i P_i}{E_i^*} + \frac{\chi_i Q_i}{E_i^*} \tag{3.16}$$

Although this method improves the voltage regulation but still it depends on the active power control. Therefore, this problem can be solved by assuming the voltage drop as a non-linear function of both P and Q as follows:

$$E_i = E_{i,\text{ref}}^* - K_i(P_i, Q_i)Q_i + \frac{\gamma_i P_i}{E_i^*} + \frac{\chi_i Q_i}{E_i^*} \tag{3.17}$$

$$K_i(P_i, Q_i) = K_{Qi} + K_{Qi}Q_i^2 + K_{Pi}P_i^2$$

This technique improves the reactive power sharing and stability margin. Although, this technique modifies the droop controller by considering the resistance and reactance of the feeder, however, small errors may cause system instability.

- **(P-Q)-ω/(P+Q)-E Droop:** To avoid the dependency of P and Q on line impedance, the authors in Ref. [55] proposed an improved droop controller that can be expressed as:

$$\omega = \omega_{\text{ref}} - K_P\left\{(P-Q)-(P_{\text{ref}} - Q_{\text{ref}})\right\} \tag{3.18}$$

$$E = E_{\text{ref}} - K_Q\left\{(P+Q)-(P_{\text{ref}} + Q_{\text{ref}})\right\} \tag{3.19}$$

From the above expressions, it can be observed that in this method both P and Q are coupled together; thus, this approach improves the performance of E-P/Q-ω controller while ensuring appropriate power-sharing with and shows a good dynamic response for resistive networks.

- **Virtual Frame Transformation:** In an MG, the impedance feeders have comparable inductive and resistive part; as a result, they always suffer from coupling effects. Thus, by using a conventional droop, accurate power-sharing cannot be achieved. Therefore, to cope with this challenge, an orthogonal transformation is used in which P and Q are transformed into a new reference frame to achieve virtual power [56]. In this method, P and Q are independent from the feeder impedance and can be given as:

$$\begin{bmatrix} P^\circ \\ Q^\circ \end{bmatrix} = \begin{bmatrix} \sin\theta & -\cos\theta \\ \cos\theta & \sin\theta \end{bmatrix} \begin{bmatrix} P \\ Q \end{bmatrix} \tag{3.20}$$

These transformed P and Q can be used in droop characteristics equations. Similarly, E and ω can be transformed in this frame by keeping the power terms as it is, thus it can be given as [57]:

$$\begin{bmatrix} \omega^\circ \\ E^\circ \end{bmatrix} = \begin{bmatrix} \sin\theta & \cos\theta \\ -\cos\theta & \sin\theta \end{bmatrix} \begin{bmatrix} \omega \\ E \end{bmatrix} \tag{3.21}$$

The droop characteristics equations with virtual terms can be written as:

$$\omega^\circ = \omega_{\text{ref}}^\circ - K_P P \tag{3.22}$$

$$E^\circ = E_{\text{ref}}^\circ - K_Q Q \tag{3.23}$$

In this method, the authors only consider linear loads, whereas the non-linear loads are not taken into consideration. Furthermore, during the process, the transformation angles must be same, any mismatch results in inaccurate transformation of E and ω.

- **Virtual Output Impedance (VOI) Controller:** In parallel-connected converters, accurate power-sharing cannot be guaranteed in a line impedance mismatch condition. Therefore, to cope with this issue, the authors in Ref. [58] proposed a solution of virtual impedance. An additional VOI control loop is incorporated with the droop controller to adjust the output impedance of the inverter in order to decouple P and Q. Based on this methodology, a resultant reference voltage can be presented as:

$$E_{\text{ref}} = E_{0i} - Z_E(s)i_{0i} \tag{3.24}$$

where E_{0i} and i_{0i} are the output voltage and current at no-load condition respectively and $Z_E(s)$ represents the VOI and for pure inductive line $Z_E(s) = L_E S$. This method shows efficient performance for linear loads however, for non-linear loads harmonic distortion is observed. The harmonics can be eliminated by using a high pass filter along with the VOI as:

$$E_{\text{ref}} = E_{0i} - \frac{L_E S}{S + \omega_c} i_{0i} \tag{3.25}$$

where ω_c is the cut-off frequency of the filter. Moreover, a properly tuned time-varying VOI has the capability to eliminate the current spikes during the plug-in operation in the MG and it can be formulated as [59]:

$$Z_E(t) = Z_f - (Z_f - Z_i)e^{-t/\tau} \tag{3.26}$$

where Z_f and Z_i presents the final and initial values of VOI and τ represents the soft start-up time constant. This controller has the capability to handle non-linear loads; however, if τ is not selected accurately, it will result in poor voltage and frequency regulation.

- **Enhanced VOI Controller:** To improve the quality of Q and reduce the harmonic effects, the authors in Ref. [60] proposed an enhanced VPI control loop. This technique compensates for the effect of feeder impedance mismatching and provides appropriate power-sharing. This controller determines the virtual impedance at selected as well as fundamental harmonic frequencies and thus eliminates the harmonics generated at PCC. Moreover, this method requires a low-bandwidth CN and also depends on the parameters of line impedance.

- **Droop-Based Synchronization:** To eliminate the line impedance mismatch and ensure accurate power-sharing, an enhanced droop controller is proposed in Ref. [61]. The proposed controller works in two operating levels, i.e., error minimization and voltage regulation. In the first level, the accuracy of reactive power sharing is improved by minimizing the error, and as a result, the voltage magnitude decreases. Moreover, a first-level operation is implemented through a low-bandwidth CN. In a second level, a deviated voltage is restored following the equation given as:

$$E_i(t) = E_{\text{ref}} - K_{Qi} Q_i(t) + \sum_{n=1}^{k} G^n E - \sum_{n=1}^{k} K_i Q_i^n \qquad (3.27)$$

$$\omega_i(t) = \omega_{\text{ref}} - K_{Pi} P_i(t) \qquad (3.28)$$

where G^n presents the input signal for restoring voltage at kth-interval and its value can either be 0 or 1; if $G^n = 1$ then it means that the voltage restoration in performed. E represents the voltage restoration constant, k shows the synchronization events count number till time t, and Q_i^n is the Q output at nth-synchronization interval. This control scheme shows efficient performance under PnP operation and shows robustness against time-varying communication delays.

- **Signal Injection Method:** To improve power-sharing, a small injection-based controller is presented in Ref. [62] as shown in Figure 3.14. In this method, every DG injects a high frequency signal into the MG that is a function of P, Q, and distortion power. The operating frequency of an injected signal (ω_q) based on the output Q of an ith-DG can be presented as:

$$\omega_{qi} = \omega_{q0} + Q_i D_{Qi} \qquad (3.29)$$

where ω_{q0} is the reference of the angular frequency and D_{Qi} presents a boost coefficient. An injection of small high frequency voltage results in small active power flow (p_{qi}) in an MG network and the output voltage of ith-DG can be adjusted as:

$$E_i = E_{\text{ref}} - p_{qi} D_{Pi} \qquad (3.30)$$

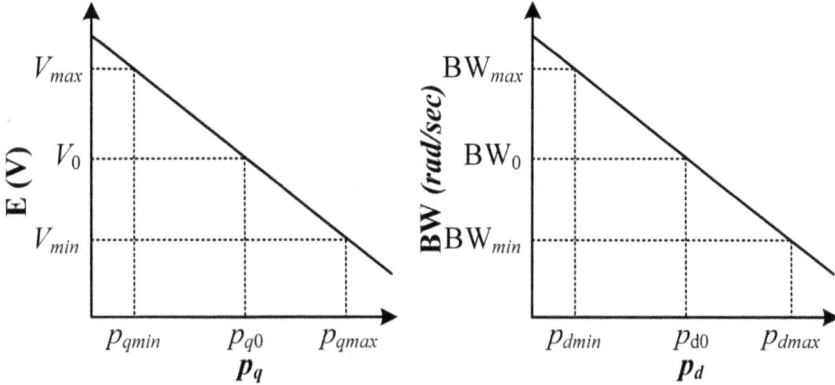

FIGURE 3.14   Droop curves with signal injection.

By using (3.30), Q/E droop is accomplished through a frequency control input. In this method, initially, the frequency of the injected signal is dropped based on distortion power. The power transferred by the injected signal (p_{di}) is used to regulate the MG and its droop characteristics equations are given as follows:

$$\omega_{di} = \omega_{d0} - K_i D_i \tag{3.31}$$

$$BW_i = BW_0 - D_{bw} p_{di} \tag{3.32}$$

where ω_{d0} is the frequency of the injected signal (voltage control input) and BW_0 presents the bandwidth of the voltage control loop.

This control technique does not depend on the variation of line impedance; therefore, it ensures accurate power-sharing and has the capability to efficiently work under linear as well as non-linear loads. Beside these numerous advantages, there are some drawbacks, such as accurate E-regulation is not guaranteed, causes harmonic current, and has high complexity.

- **Droop with Derivative and Integral Compensation:** An adaptive droop controller proposed in Ref. [63] is feasible for IS as well as GC mode. In this method, a derivative term is included in the active power droop equation to improve the dynamics of the power loop in IS mode of operation. On the contrary, an integral term is employed in the Q droop equation to enhance the power factor during GC mode. The P and Q droop equations using this methodology can be expressed as:

$$\omega = K_P P + K_P \frac{dP}{dt} \tag{3.33}$$

$$E = K_Q Q + K_Q \int Q \, dt \tag{3.34}$$

- **Droop based on Common Bus Voltage:** To ensure accurate power-sharing among the parallel-connected DGs, the authors in Refs. [64,65] proposed an improved droop controller. In this method, an integral controller is used to regulate the reactive power sharing by tracking the common bus voltage as:

$$\begin{cases} E_i = k_{qi} \int (E_{\text{ref}} - E_{\text{com}}) dt \\ E_{\text{ref}} = E_{oi} - K_{Qi} Q \end{cases} \tag{3.35}$$

where k_{qi} presents an integral gain, E_{com} and E_{oi} represents the remote and local reference signals respectively, and the reference output voltage is denoted by E_{ref}. Under steady-state condition, $E_{\text{ref}} = E_{\text{com}}$ as a result the input becomes zero and the related Q can be calculated as:

$$Q_i = \frac{E_{oi} - E_{\text{com}}}{K_{Qi}} \tag{3.36}$$

From the above expression, it can be concluded that Q_i is the linear function of E_{oi}, therefore, it can be said that under steady-state the reactive power is not affected by the parameters of the MG network. For the resistive MG, the authors in Ref. [65] follow a similar procedure and the resultant active power control can be expressed as:

$$E_i = \int \left[k_E (E_{\text{ref}} - E_{\text{com}}) k_q P_i \right] dt \tag{3.37}$$

where k_E is the gain of amplifier. This controller proportionally share the load among the DGs and is independent of the disturbances, component mismatch, and MG parameters.

A concise review of the above-discussed droop control-based power-sharing strategies is summarized in Table 3.1. It includes their implementation complexity, their dependency on system parameters, the accuracy of reactive power sharing, the type of load analysed, and their dynamic speed.

TABLE 3.1 Review of Different Droop-Based Control Schemes

Ref.	Controller	Implementation	Reactive Power Sharing	System Dependency	Dynamic Response	Line	Type of Load
[39]	$P\text{-}\omega/Q\text{-}E$	Easy	Inaccurate	Yes	Slow	MV	Linear
[44]	Adaptive $P\text{-}\omega$/adaptive $E\text{-}Q$ droop	Complex	Accurate	Yes	Improved	MV	Linear, Non-linear
[49]	$P\text{-}E/Q\text{-}\omega$	Easy	Inaccurate	Yes	Slow	LV	Linear
[51]	$P\text{-}\delta/Q\text{-}E$	Complex	Inaccurate	Yes	Slow	MV, LV	Linear, Non-linear
[53]	Improved droop	Complex	Accurate	No	Improved	LV	Linear, Non-linear
[54]	Adaptive $P\text{-}\delta/E\text{-}Q$	Complex	Inaccurate	Yes	Slow	MV, LV	Linear, Non-linear
[55]	$(P\text{-}Q)\text{-}\omega/(P\text{+}Q)\text{-}E$	Complex	Accurate	No	Improved	LV	Linear
[56]	Virtual frame transformation	Easy	Inaccurate	Yes	Improved	LV	Linear
[58]	VOI	Easy	Accurate	No	Slow	MV, LV	Linear, Non-linear
[60]	Enhanced VOI	Complex	Accurate	Yes	Improved	MV, LV	Linear, Non-linear
[61]	Droop-based synchronization	Complex	Accurate	Yes	Improved	MV, LV	Linear, Non-linear
[62]	Signal injection	Complex	Inaccurate	No	Slow	MV, LV	Linear, Non-linear
[63]	Droop with derivative and integral compensation	Easy	Inaccurate	Yes	Improved	LV	Linear
[64]	Droop-based on common bus voltage	Complex	Accurate	No	Improved	MV, LV	Linear, Non-linear

3.2.2.2 Non-Droop-Based Control Methods

Non-droop-based control techniques (such as centralized control, distributed control, and master-slave) provide efficient performance in terms of voltage regulation and reactive power sharing. As compared to the droop-based control methods, these controllers restore the frequency and voltage close to their reference values. However, these controllers require a CN between the inverters that results in cost increment.

- **Centralized Control:** A generalized schematic of a centralized control structure that involves current division modules and synchronization signals is shown in Figure 3.15 [66]. Every module or DG unit has Phase Locked Loop (PLL) that is responsible for eliminating the mismatch between the synchronization signal and the output voltage (phase and frequency). Furthermore, current division units are used to detect load consumption and provide the current reference that is needed to be drawn from each DG. In inverter-interfaced MGs, the N-DGs (same capacity) share the current according to the following expression:

$$i_g = i_k / N \tag{3.38}$$

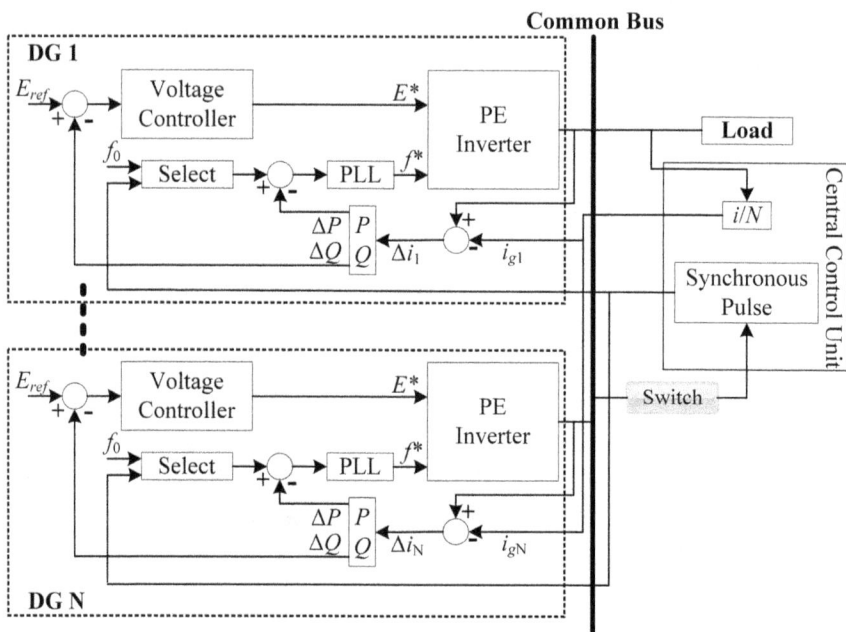

FIGURE 3.15   Schematic of centralized controller.

where i_k is the output current of inverter and $k=1,2,...N$. Every inverter of the DG measures the current error Δi_k by subtracting the reference current $i_{ref} = i_g$ from i_k. As the frequency and phase of the inverters are synchronized through synchronization signal, hence, Δi_k arises due to variation in the magnitude of the output voltages. A change in Q i.e. ΔQ and the reference voltage i.e. E_{ref} decide the input voltage value of the inverter $E*$. Moreover, a change in P, i.e. ΔP and a frequency set point i.e. f_0 decide the frequency $f*$ for the inverter.

This controller ensures accurate current in steady-state as well as in dynamic. However, it is based on one centralized controller, which reduces system redundancy and scalability. It requires the high-bandwidth CN to send information from the controller to inverters thus results in high cost. Moreover, due to its high dependency on CN, the reliability of the system is reduced because of single-point failure.

- **Distributed Control:** In a distributed control scheme, no central controller is involved; instead, the controller is implemented locally at each DG, and the exchange of information between them is performed through communication lines. Moreover, every DG in an MG is connected with its neighbouring DGs through CN in order to share the key parameters. Usually, a low-bandwidth CN is used for the exchange of information with each other and can be deployed in IS mode as well as in GC mode. A general schematic of the distributed controller that is characterized by average current sharing is presented in Figure 3.16 [67]. In this control method, a current sharing bus is required for reference voltage synchronization and the sharing of average current. Moreover, to ensure accurate tracking of the reference current and to ensure that the reference current is equally shared among all the parallel-connected DGs an additional current control is also used.

From Figure 3.16, it can be seen that initially the current error i_{eN} is transformed from abc reference frame to dq reference frame. Then the frequency and amplitude of the output voltage are regulated through current regulators. The output current of DGs i_{ok} $(k=1,2,...,N)$ is averaged as follows:

$$i_{avg} = \frac{1}{N} \sum_{k=1}^{N} i_{ok} \qquad (3.39)$$

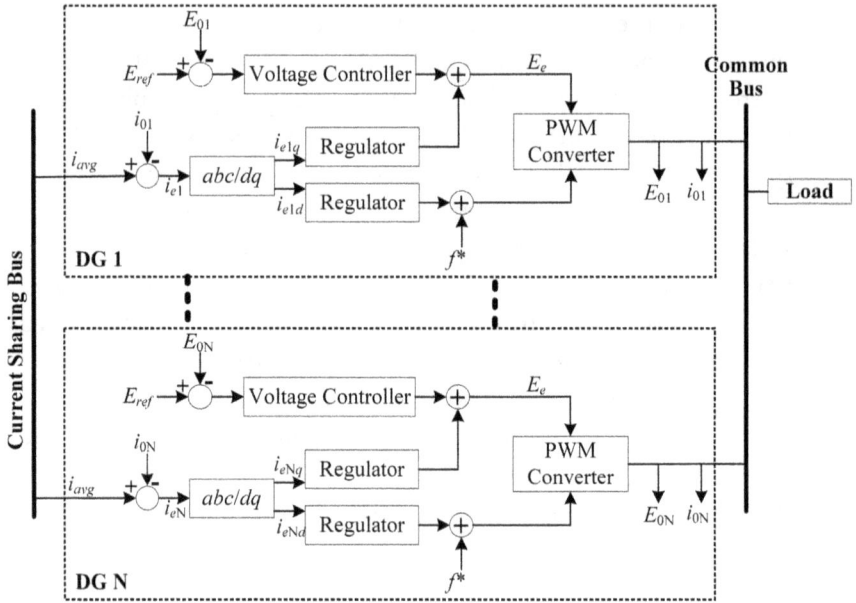

FIGURE 3.16 Schematic of distributed controller.

The output current i_{oN} and voltage E_{oN} are feedback and compared them with the average current (provided by current sharing bus) and reference voltage, respectively.

In a distributed control, every DG requires the information of its adjacent DGs; therefore, compared to a centralized control, it requires lower bandwidth CN. Moreover, in case of failure of one DG does not affect the performance and parallel operation of the other DGs in an MG.

- **Master-Slave Control:** In a master-slave control technique, one module acts as a master while the other modules act as slaves. This type of control architecture can be seen in traditional power systems, where the slack bus regulates the frequency and voltage of the system while all the other buses fulfil the demands of the load. Thus encouraged by this idea, the same methodology is used in an MG where one DG unit (a DG close to the PCC is usually chosen a master DG) acts as master DG while other DG units serve as slaves, as shown in Figure 3.17 [68]. A voltage controller in a master DG unit regulates the output voltage of the inverter or DG by tracking the reference voltage (E_{ref}). The controller output voltage (E_e) directs the PWM of the master DG and is

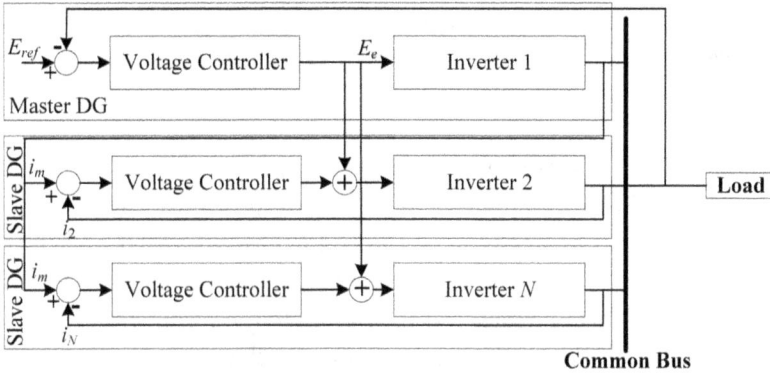

FIGURE 3.17   Schematic of master-slave controller.

then fed to slave inverters in order to regulate the voltage. Moreover, the master's inverter output current (i_m) acts as a reference current $(i_m = i_{ref})$ for all slave inverters. The slave inverters track the reference provided by master inverter and ensures accurate and proportionate power-sharing. This control method is simple to implement, avoid the use of PLL signals as all the DGs are communicated with the master DG, ensure accurate power sharing, and enables PnP functionality. However, beside these numerous advantages, one of the major disadvantages of this system is that if a master DG fails to operate, it may cause the system instability. Moreover, as the master inverter operated in voltage control mode therefore during the transition phase the output current may overshoot as it is not controlled.

3.3 SECONDARY CONTROL LEVEL

To overcome the limitations of the primary controller, an SC level is designed with the key objective of eliminating the voltage and frequency deviations of the DGs caused by the primary control level while ensuring appropriate power-sharing. Moreover, harmonics compensation, voltage unbalancing, power quality enhancement, reconnection of MG with UG, etc., are few of the control objectives performed by the SC level [69]. A secondary controller is designed to operate at a slower time frame compared to the primary control level to facilitate the steady-state-attainment of primary controller before SC updates its points. SC level works primarily in cooperation with the communication network and is well suited for optimization problems such as economic dispatch, day scheduling and power-sharing optimization along with restoration and synchronization problems. Hence, based on the communication network, the SC structure can be categorized as a centralized, decentralized, and distributed control architecture.

3.3.1 Centralized Secondary Controller

In a centralized control architecture, a communication system is used to connect every DG to a Microgrid Central Controller (MGCC) [70]. An MGCC initially collects the measured data, makes the mandatory calculations, and based on the data sends the signals to the secondary controllers to perform the required tasks. In this control architecture, a specific node in an MG is selected as the control bus, where the voltage and frequency are measured and estimated [58]. A point in a sensitive load bus can be selected as the control bus, but generally, a PCC is selected as the control bus [71]. A generalized schematic of this control approach is presented in Figure 3.18. From Figure 3.18, it can be seen that the MGCC is connected to every DG and the information is sent to the SC to perform its control actions. In this case, the CN follows a star structure, i.e., there are no communication links among the DGs, only the communication links between DGs and MGCC are required. If it is needed to send information from one DG to another, then the information will first be sent to the MGCC, and then from MGCC, the information will be forwarded to the related DG.

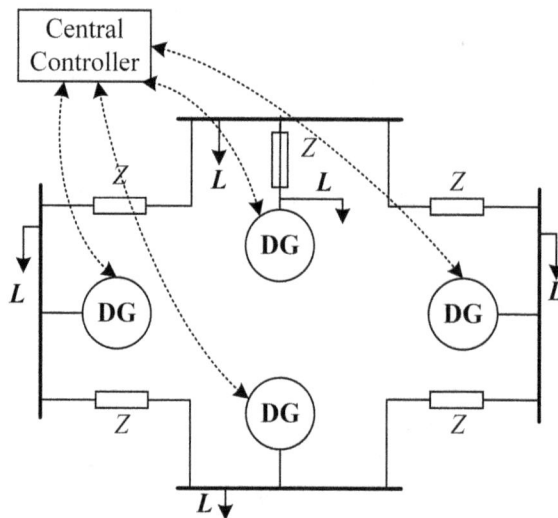

FIGURE 3.18 Schematic of centralized secondary controller.

A centralized controller is proposed in Ref. [58], in which the DG's frequency and voltage are compared with their reference values and their corresponding error $\delta\omega$ and δE are proceeded via PI controller and sent to the primary control of DG. The error term can be given as:

$$\delta\omega_i = K_{p\omega}(\omega_{\text{ref}} - \omega_i) + K_{I\omega} \int (\omega_{\text{ref}} - \omega_i)dt + \Delta\omega_{\text{syn}} \qquad (3.40)$$

$$\delta E_i = K_{pE}(E_{\text{ref}} - E_i) + K_{IE} \int (E_{\text{ref}} - E_i)dt \qquad (3.41)$$

where $K_{p\omega}$ and K_{pE} are the proportional gain of PI controller while the integral gain are presented as $K_{I\omega}$ and K_{IE}, $\Delta\omega_{\text{syn}}$ is the synchronization input whose value is zero in IS mode, i.e., $\Delta\omega_{\text{syn}} = 0$. $\delta\omega$ and δE are the maximum allowable frequency and voltage deviation.

The authors in Ref. [72] proposed a potential function-based centralized SC in which a potential functional for every DG is computed by the control unit that contains information about measurements, control input, and constraints. A potential function for ith-DG can be expressed as:

$$p_k(x_k) = w_c \sum_{i=1}^{m_c} \text{pp}_i^c(x_k) + w_u \sum_{i=1}^{n_u} \text{pp}_i^u(x_k) + w_g \text{pp}_i^g(x_k) \qquad (3.42)$$

where p_k denotes the potential function, x_k is the measurement (interested measurement, e.g. P, Q, f, E), pp_i^c, pp_i^u, and pp_i^g are the partial potential function for constraints, measurements, and set point respectively, and their corresponding weights can be presented as w_c, w_u, and w_g respectively.

This architecture requires a widespread CN between the DGs and MGCC, which makes it suitable to ensure a high-quality voltage at a specific node. Moreover, it is also useful to control and monitor different aspects of MG as a single entity. However, in scenarios when MG is dispersed in complex geographical areas, it may result in high communication infrastructure costs. Moreover, during PnP functionality, the scalability of the MG is one of the main drawbacks of this control architecture. As the addition of a new DG requires an update the settings of MGCC. Furthermore, another major drawback of this approach is the dependency of the whole system on one controller. In case of failure or malfunctioning of MGCC, it will affect the performance of entire system and may lead to system instability.

Different controllers such as PI and MPC along with their merits, application, and demerits employed in a centralized control architecture are summarized in Table 3.2.

TABLE 3.2 Different Centralized Controllers' Overview

Ref.	Controller	Merits	Demerits	Applications
[72]	Potential function-based controller	Easy implementation Flexible High controllability	Complex CN Information retrieval and embedding are difficult	Power management Voltage control
[73]	PI	Flexible Operation Easy Implementation	Single point failure	Frequency regulation Voltage regulation
[74]	MPC	Simple control method without modulation Constraint handling capability Improve system performance	Not suitable for large load disturbance (Linear MPC)	Power management Frequency and voltage control
[75]	Hybrid MPC	Fast computation time Improve system reliability and power quality	Poor voltage regulation	Coordinates parallel operations of DGs
[76]	Coordination control	Scalable nature Equalizes the SOC of ESS Storage capacity independency	Negative weight factor may cause system instability	Coordinated operation of ESS and DGs for frequency and voltage regulation
[77]	Estimation-based controller	Improve the frequency stability of the system by load shedding	Complex Implementation	In an IS MG improves the frequency stability
[78]	Compensation-based controller	Use low-bandwidth CN Reduces the voltage harmonics	High complexity in harmonic signal sharing	Voltage harmonic compensation

3.3.2 Distributed Secondary Controller

In a centralized control architecture, all the control responsibilities come on a single centralized controller that is very feasible and reliable. Today, more and more RESs are integrated into the MGs, which makes the control structure more complex. Moreover, with the passage of time, more control objectives are also defined. So, in the presence of a highly penetrated MG, it is very difficult for the centralized controller to control all the objectives.

Moreover, due to the usage of a single centralized controller, the reliability and flexibility of the system also decrease, as if a centralized controller fails to operate. Therefore, to increase the reliability and flexibility of the system, the concept of a distributed SC arises. In a distributed SC, every DG is connected to a separate controller that performs the control actions. Moreover, every DG is also connected to its adjacent neighbouring DG and shares information with them through CN. This avoids the problem of single point failure as experienced in MGCC [79]. It should be kept in mind that in some applications, even with the SC, there is a need for MGCC, such as coordination among the DGs during the start processes in case of blackouts, etc.

A generalized schematic of the distributed SC is presented in Figure 3.19. From Figure 3.19, it can be seen that there is separate controller for every

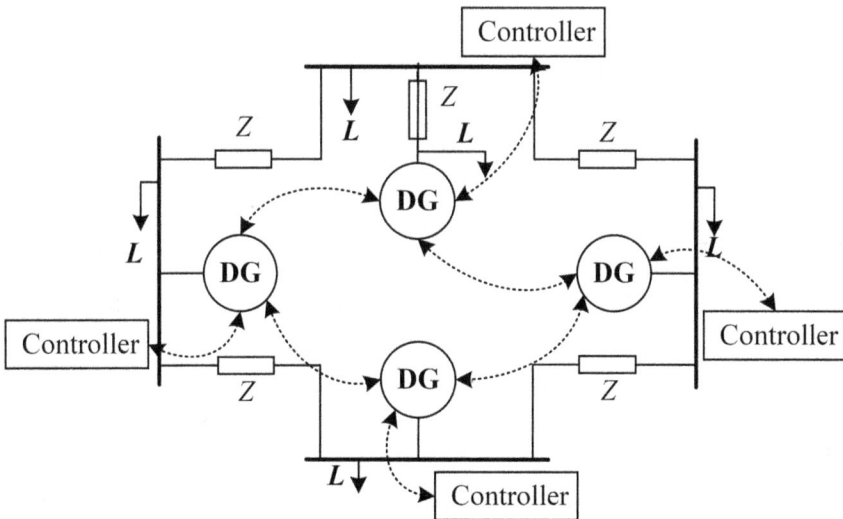

FIGURE 3.19 Schematic of distributed secondary controller.

DG and all the DGs are connected with the neighbouring DG that share the information through spare communication network. The distributed SC can be categorized into different groups based on the control calculation techniques. These are discussed below in detail.

3.3.2.1 Consensus-Based Method

In this method, every DG uses the measurements of its neighbouring DG units or agents based on Multi-Agent System (MAS) theory [80]. A communication system in a consensus technique is based on the neighbour to neighbour communication; as a result, high robustness can be attained. A consensus technique can be presented as:

$$u_i^P = k_C \left(\int \left(\alpha_{Ci} e_i + \beta_{Ci} \varepsilon_i \right) dt - \omega_i \right) \tag{3.43}$$

$$e_i = \gamma_{Ci} \left(\omega_{ref} - \omega_i \right) + \sum_{j=1}^{m} \left(\omega_j - \omega_i \right) \tag{3.44}$$

$$\varepsilon_i = \sum_{j=1}^{m} \left(u_j^P - u_i^P \right) \tag{3.45}$$

where k_C, α_{Ci}, β_{Ci}, and γ_{Ci} are the control parameters of the consensus method. In (3.45), a tracking frequency error (e_i) formula consists of two terms, i.e., a first term $(\omega_{ref} - \omega_i)$ is related to the frequency error at the local inverter while a second term $\sum_{j=1}^{m} (\omega_j - \omega_i)$ is the averaging of the frequency deviation that occurs between the m-neighbouring DGs and the local inverter. Moreover, a term ε_i presents the averaging terms of secondary layer between m-neighbours and the local inverter.

Numerous advancements have been made in the consensus technique such as the authors in Ref. [81] describe a consensus technique in which a voltage control based on the averaging technique is combined with a frequency control that is based on the consensus technique. In this technique, two terms are used to regulate the voltage, i.e., one term considers reactive power error in correspondence with the other DGs while the second term considers the E-error; and can be given as:

$$u_i^E = \beta_{Ci} \int \sum_{j=1}^{n} a_{ij} \left(\frac{Q_j}{Q_{j\,\text{ref}}} - \frac{Q_i}{Q_{i\,\text{ref}}} \right) + \gamma_{Ci} \int (E_{\text{ref}} - E_i) dt \qquad (3.46)$$

where a_{ij} is the related adjacency matrix which describes the communication link the ith-agent and jth-agent in MAS.

An example of the five agent systems connected to each other through a CN is presented in Figure 3.20. In Figure 3.20, the agents or DGs are presented by circles while the dotted lines show the cyber-links. An adjacency matrix related to the Figure 3.20 can be given as:

$$A = \begin{bmatrix} 0 & a_{12} & 0 & a_{14} \\ a_{21} & 0 & a_{23} & 0 \\ 0 & a_{31} & 0 & a_{34} \\ a_{41} & 0 & a_{43} & 0 \end{bmatrix} \qquad (3.47)$$

From (3.47), it can be say that $a_{ij} > 0$ if there it cyber link present between agent i and agent j, whereas, if there is no communication link then $a_{ij} = 0$. This matrix can also be used to present the unidirectional cyber connection where $a_{ij} \neq a_{ji}$.

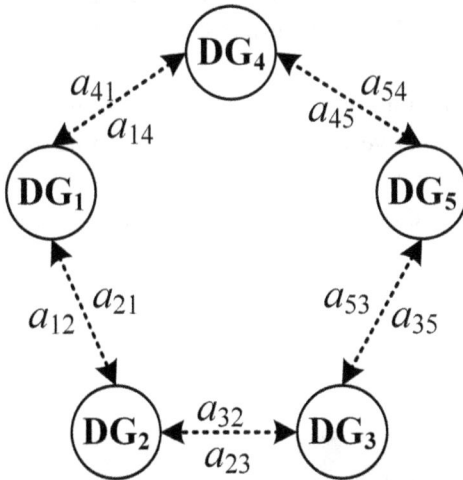

FIGURE 3.20 Multi-agent system with sparse cyber-links.

A consensus-based method has numerous advantages, such as a reduction in CN infrastructure, high flexibility, and the capability to adopt different MG topologies. However, besides these numerous advantages, a major drawback of this method is that as the number of DGs increases, the CN infrastructure complexity also increases, which may affect the performance of the system in case of any delays or cyber-attacks [82].

3.3.2.2 Averaging Method

In this method, every converter of the DG computes its secondary control terms using the measured values of all or some of the other converters of the DGs. As a result, a global average value is formed that follows a control reference, thus resulting in fast convergence of the system. In this method, a local frequency is estimated and sent to all the inverters of the DGs in order to calculate the average frequency, i.e., ω_{avg}. Hence, based on this, the secondary control term can be calculated as:

$$u_i^\omega = k_p \left(\omega_{ref} - \omega_{avg} \right) + k_i \int \left(\omega_{ref} - \omega_{avg} \right) dt \tag{3.48}$$

$$\omega_{avg} = \frac{1}{n} \sum_{i=1}^{n} \omega_i \tag{3.49}$$

where k_p and k_i are the proportional and integral gains of the PI controller while the number of DGs or inverters connected to the MG are denoted by n [83].

As the voltage is not constant at the nodes, it is therefore difficult to achieve accurate reactive power sharing. Therefore, an averaging controller is designed that not only considers E-regulation but also reactive power sharing is taken into consideration; and can be expressed according to [84] as:

$$u_i^E = u_i^{E_{avg}} + u_i^{Q_{avg}} \tag{3.50}$$

$$u_i^{E_{avg}} = \alpha_p \left(E_{ref} - E_{avg} \right) + \alpha_i \int \left(E_{ref} - E_{avg} \right) dt \tag{3.51}$$

$$u_i^{Q_{avg}} = \beta_p \left(Q_{ref} - Q_{avg} \right) + \beta_i \int \left(Q_{ref} - Q_{avg} \right) dt \tag{3.52}$$

where α_p, α_i, β_p, and β_i presents the control parameters of PI controller while E_{avg} and Q_{avg} denotes the average voltage and reactive power respectively.

3.3.2.3 Event-Triggered Method

In a consensus technique, data is measured through CN, and the distributed controller has the responsibility to control the signals. However, this technique is not very feasible due to the continuous utilization of the CN; thus; as a result, numerous issues arise, such as packet loss and delays that may affect the system's reliability and performance. Hence, to cope with these challenges, an event-triggered technique is a feasible solution to maintain the controllability in MAS. Moreover, this method also decreases the excessive use of CN; as a result, computational burden will also be reduced [85].

This method is widely used in MG as it decreases the exchange of information among the DGs. Furthermore, a consensus technique is used to examine the controllability of the event trigger that can be divided into three types: (a) Event triggered, (b) time triggered, and (c) self-triggered sampling methods. In the event-triggered technique, different from the continuous transfer of signal, a signal is only sent to the secondary controller in case of an occurrence of an event. Moreover, in a sampling process, it can be considered as an introductory feedback that requires an enduring checking of the states to accurately determine the present performance. In a second type of event-triggered technique, i.e., in a time-triggered method, a clock is used to drive the control [86]. In a self-triggered technique, a current state is not only used to compute the input control signal of the system, but it is also used for the re-computation of the control law a next time. In this method, an enduring monitoring or checking of the state is not required, just like the event-triggered technique [87].

3.4 TERTIARY CONTROL LEVEL

It is the highest control level in a hierarchical structure and has the lowest operation speed among all the control levels. It is mostly related to the optimum and economic operation of MG and manages power transmission. In a normal operating mode, power transmission is regulated by adjusting the frequency and voltage of the DG sources. In this control level, initially, a controller detects the output power of the MG and then compares the measured values with the nominal references. After the power comparison, the frequency (ω_{ref}) and voltage (E_{ref}) are obtained as follows:

$$\omega_{\text{ref}} = k_{pP}\left(P_G^{\text{ref}} - P_G\right) + k_{iP}\int\left(P_G^{\text{ref}} - P_G\right)dt \tag{3.53}$$

$$E_{\text{ref}} = k_{pQ}\left(Q_G^{\text{ref}} - Q_G\right) + k_{iQ}\int\left(Q_G^{\text{ref}} - Q_G\right)dt \tag{3.54}$$

where k_{pP}, k_{iP}, k_{pQ}, and k_{iQ} are the proportional and integral control coefficients, ω_{ref} and E_{ref} are the frequency and voltage reference values for the SC.

To provide an optimal economic operation of the MG, tertiary controller must ensure that all DGs must be operated at equal marginal cost. This can be achieved by using an advanced and intelligent algorithm at a tertiary level that describes the optimum cost coefficient. In this process, to initiate the algorithm, the starting power values for the ith-DG and jth-DG are selected arbitrarily by the cost coefficient. Thus, the iterations start and continue to operate until the initial values reach the optimum values. Moreover, the output power of the DGs is continuously measured until the required power values are attained. Once, the required values are attained, a controller will send a command that forces the DGs to generate the desired output power [45]. This searching phenomena is performed on every pair of DG, and thus all the DGs in the MG are forced to operate at the optimal value. Numerous algorithms based on metaheuristics, game theory, etc., are used to facilitate CN in order to obtain the measurement results more easily [88].

3.5 CONCLUSION

In this chapter, a detailed discussion of the hierarchical control structure of an MG is provided. A primary control level that consists of an inverter output control loop and a power-sharing control loop is elaborated in detail. Different control reference frames used in the inverter output control loop are discussed. Moreover, a power sharing controller is further categorized into droop-based and non-droop-based control strategies, and provides a detailed description of these controllers. Furthermore, a detailed description of the centralized and distributed SC level is provided, and different controllers and methods used in these control architectures are explained. Finally, the characteristics and working operation of tertiary control level are presented.

3.6 EXERCISES

1. Explain the typical control structure of microgrid.

2. Describe the similarities and differences between droop and non-droop-based control methods.

3. Compare the fundamentals of DC and AC droop control methods and explain the difference.

4. Referring to Section 3.2.1.3, use different control strategies discussed in Section 3.2.1.2 in the outer voltage and inner current control loops and make a comparative analysis of these controllers.

REFERENCES

1. Ali, Hossam, Gaber Magdy, and Dianguo Xu. "A new optimal robust controller for frequency stability of interconnected hybrid microgrids considering non-inertia sources and uncertainties." *International Journal of Electrical Power & Energy Systems* 128 (2021): 106651.
2. Som, Shreyasi, Souradip De, Saikat Chakrabarti, Soumya Ranjan Sahoo, and Arindam Ghosh. "A robust controller for battery energy storage system of an islanded AC microgrid." *IEEE Transactions on Industrial Informatics* 18, no. 1 (2021): 207–218.
3. Guerrero, Josep M., Jos Matas, Luis Garcia de Vicuna, Miguel Castilla, and Jaume Miret. "Decentralized control for parallel operation of distributed generation inverters using resistive output impedance." *IEEE Transactions on Industrial Electronics* 54, no. 2 (2007): 994–1004.
4. Lopes, J. A. Peças, Carlos L. Moreira, and A. G. Madureira. "Defining control strategies for microgrids islanded operation." *IEEE Transactions on Power Systems* 21, no. 2 (2006): 916–924.
5. Ahmed, Kafeel, Mehdi Seyedmahmoudian, Saad Mekhilef, N. M. Mubarak, and Alex Stojcevski. "A review on primary and secondary controls of inverter-interfaced microgrid." *Journal of Modern Power Systems and Clean Energy* 9, no. 5 (2020): 969–985.
6. Ali Khan, Muhammad Yasir, Haoming Liu, Zhihao Yang, and Xiaoling Yuan. "A comprehensive review on grid connected photovoltaic inverters, their modulation techniques, and control strategies." *Energies* 13, no. 16 (2020): 4185.
7. Zeb, Kamran, Imran Khan, Waqar Uddin, Muhammad Adil Khan, P. Sathishkumar, Tiago Davi Curi Busarello, Iftikhar Ahmad, and H. J. Kim. "A review on recent advances and future trends of transformerless inverter structures for single-phase grid-connected photovoltaic systems." *Energies* 11, no. 8 (2018): 1968.

8. Zhang, Li, Kai Sun, Yan Xing, Lanlan Feng, and Hongjuan Ge. "A modular grid-connected photovoltaic generation system based on DC bus." *IEEE Transactions on Power Electronics* 26, no. 2 (2010): 523–531.

9. Sathiyanarayanan, T., and Sukumar Mishra. "Synchronous reference frame theory based model predictive control for grid connected photovoltaic systems." *IFAC-PapersOnLine* 49, no. 1 (2016): 766–771.

10. Huang, Qicheng, and Kaushik Rajashekara. "A unified selective harmonic compensation strategy using DG-interfacing inverter in both grid-connected and islanded microgrid." In *2017 IEEE Energy Conversion Congress and Exposition (ECCE)*, pp. 1588–1593. IEEE, Cincinnati, OH, USA, 2017.

11. Zolfaghari, Mahdi, Ali Asghar Khodadoost Arani, G. B. Gharehpetian, and Mehrdad Abedi. "A fractional order proportional-integral controller design to improve load sharing between DGs in microgrid." In *2016 Smart Grids Conference (SGC)*, pp. 1–5. IEEE, Kerman, Iran, 2016.

12. Yuan, Xiaoming, Willi Merk, Herbert Stemmler, and Jost Allmeling. "Stationary-frame generalized integrators for current control of active power filters with zero steady-state error for current harmonics of concern under unbalanced and distorted operating conditions." *IEEE Transactions on Industry Applications* 38, no. 2 (2002): 523–532.

13. Lenwari, Wanchak. "Optimized design of modified proportional-resonant controller for current control of active filters." In *2013 IEEE International Conference on Industrial Technology (ICIT)*, pp. 894–899. IEEE, Cape Town, South Africa, 2013.

14. Teodorescu, Remus, Frede Blaabjerg, Marco Liserre, and P. Chiang Loh. "Proportional-resonant controllers and filters for grid-connected voltage-source converters." *IEE Proceedings-Electric Power Applications* 153, no. 5 (2006): 750–762.

15. Vidal, Ana, Francisco D. Freijedo, Alejandro G. Yepes, Pablo Fernandez-Comesana, Jano Malvar, Óscar López, and Jesús Doval-Gandoy. "Assessment and optimization of the transient response of proportional-resonant current controllers for distributed power generation systems." *IEEE Transactions on Industrial Electronics* 60, no. 4 (2012): 1367–1383.

16. Lim, Jae Sik, Changreung Park, Jungho Han, and Young Il Lee. "Robust tracking control of a three-phase DC–AC inverter for UPS applications." *IEEE Transactions on Industrial Electronics* 61, no. 8 (2013): 4142–4151.

17. Buso, Simone, Tommaso Caldognetto, and Qing Liu. "Analysis and experimental characterization of a large-bandwidth triple-loop controller for grid-tied inverters." *IEEE Transactions on Power Electronics* 34, no. 2 (2018): 1936–1949.

18. Falkowski, Piotr, and Andrzej Sikorski. "Finite control set model predictive control for grid-connected AC–DC converters with LCL filter." *IEEE Transactions on Industrial Electronics* 65, no. 4 (2017): 2844–2852.

19. Xia, Changliang, Tao Liu, Tingna Shi, and Zhanfeng Song. "A simplified finite-control-set model-predictive control for power converters." *IEEE Transactions on Industrial Informatics* 10, no. 2 (2013): 991–1002.

20. Cortés, Patricio, Marian P. Kazmierkowski, Ralph M. Kennel, Daniel E. Quevedo, and José Rodríguez. "Predictive control in power electronics and drives." *IEEE Transactions on Industrial Electronics* 55, no. 12 (2008): 4312–4324.

21. Chhabra, Mohit, and Frank Barnes. "Robust current controller design using mu-synthesis for grid-connected three phase inverter." In *2014 IEEE 40th Photovoltaic Specialist Conference (PVSC)*, pp. 1413–1418. IEEE, Denver, CO, USA, 2014.

22. Bahrampour, Elham, Mohammad Hassan Asemani, Farid Badfar, and Navid Vafamand. "High-performance robust grid-connected power systems." In *2019 6th International Conference on Control, Instrumentation and Automation (ICCIA)*, pp. 1–6. IEEE, Sanandaj, Iran, 2019.

23. Wu, Fengjiang, Fan Feng, Linsong Luo, Jiandong Duan, and Li Sun. "Sampling period online adjusting-based hysteresis current control without band with constant switching frequency." *IEEE Transactions on Industrial Electronics* 62, no. 1 (2014): 270–277.

24. Dey, Anubrata, P. P. Rajeevan, Rijil Ramchand, Kallarackal Mathew, and Kumarukuttan Gopakumar. "A space-vector-based hysteresis current controller for a general n-level inverter-fed drive with nearly constant switching frequency control." *IEEE Transactions on Industrial Electronics* 60, no. 5 (2012): 1989–1998.

25. Alarcón-Gallo, Eduardo, Luís García de Vicuña, Miguel Castilla, Jaume Miret, José Matas, and Antonio Camacho. "Decoupled sliding mode control for three-phase LCL VSI operating at fixed switching frequency." In *2012 IEEE International Symposium on Industrial Electronics*, pp. 1572–1578. IEEE, Hangzhou, China, 2012.

26. Espi, Jose M., Jaime Castello, Rafael Garcia-Gil, Gabriel Garcera, and Emilio Figueres. "An adaptive robust predictive current control for three-phase grid-connected inverters." *IEEE Transactions on Industrial Electronics* 58, no. 8 (2010): 3537–3546.

27. Roy, Tushar Kanti., Md Ferdous Pervej, and Farjana Khanam Tumpa. "Adaptive controller Design for grid current regulation of a CSI based PV system." In *2016 2nd International Conference on Electrical, Computer & Telecommunication Engineering (ICECTE)*, pp. 1–4. IEEE, Rajshahi, Bangladesh, 2016.

28. Kumar, Nayan, Tapas Kumar Saha, and Jayati Dey. "Sliding-mode control of PWM dual inverter-based grid-connected PV system: Modeling and performance analysis." *IEEE Journal of Emerging and Selected Topics in Power Electronics* 4, no. 2 (2015): 435–444.

29. Fei, Juntao, and Yunkai Zhu. "Adaptive fuzzy sliding control of single-phase PV grid-connected inverter." *Plos One* 12, no. 8 (2017): e0182916.

30. Hornik, Tomas, and Qing-Chang Zhong. "A current-control strategy for voltage-source inverters in microgrids based on H and repetitive control." *IEEE Transactions on Power Electronics* 26, no. 3 (2010): 943–952.

31. Zhong, Qing-Chang, and Tomas Hornik. "Cascaded current–voltage control to improve the power quality for a grid-connected inverter with a local load." *IEEE Transactions on Industrial Electronics* 60, no. 4 (2012): 1344–1355.

32. Yatak, Meral Ozarslan, and Omer Faruk Bay. "Fuzzy control of a grid connected three phase two stage photovoltaic system." In *2011 International Conference on Power Engineering, Energy and Electrical Drives*, pp. 1–6. IEEE, Malaga, Spain, 2011.

33. Thao, Nguyen Gia Minh, and Kenko Uchida. "Control the photovoltaic grid-connected system using fuzzy logic and backstepping approach." In *2013 9th Asian Control Conference (ASCC)*, pp. 1–8. IEEE, Istanbul, Turkey, 2013.

34. Khan, Muhammad Yasir Ali, Haoming Liu, Salman Habib, Danish Khan, and Xiaoling Yuan. "Design and performance evaluation of a step-up DC–DC converter with dual loop controllers for two stages grid connected PV inverter." *Sustainability* 14, no. 2 (2022): 811.

35. Srivastava, Sumit, Guguloth Ravi, Pushkar Tripathi, Leena H. Roy, Kuldeep Sahay, and Sarita Dongre. "Droop control based control technique and advancements for microgrid stability-A review." In *2023 IEEE Renewable Energy and Sustainable E-Mobility Conference (RESEM)*, pp. 1–6. IEEE, Bhopal, India, 2023.

36. Mohammed, Amira, Shady S. Refaat, Sertac Bayhan, and Haithem Abu-Rub. "AC microgrid control and management strategies: Evaluation and review." *IEEE Power Electronics Magazine* 6, no. 2 (2019): 18–31.

37. Roslan, Muhammad Farhan., Mahammad Abdul Hannan, Pin Jern Ker, and Mohammad Nasir Uddin. "Microgrid control methods toward achieving sustainable energy management." *Applied Energy* 240 (2019): 583–607.

38. Gao, Fang, and M. Reza Iravani. "A control strategy for a distributed generation unit in grid-connected and autonomous modes of operation." *IEEE Transactions on Power Delivery* 23, no. 2 (2008): 850–859.

39. Chandorkar, Mukul C., Deepakraj M. Divan, and Rambabu Adapa. "Control of parallel connected inverters in standalone AC supply systems." *IEEE Transactions on Industry Applications* 29, no. 1 (1993): 136–143.

40. Kan, Zhizhong, Zhongnan Guo, Chunjiang Zhang, and Xiaomai Meng. "Research on droop control of inverter interface in autonomous microgrid." In *2014 International Power Electronics and Application Conference and Exposition*, pp. 195–199. IEEE, Shanghai, China, 2014.

41. Pegueroles-Queralt, Jordi, Fernando Bianchi, and Oriol Gomis-Bellmunt. "Optimal droop control for voltage source converters in islanded microgrids." *IFAC Proceedings* 45, no. 21 (2012): 566–571.

42. Olivares, Daniel E., Ali Mehrizi-Sani, Amir H. Etemadi, Claudio A. Cañizares, Reza Iravani, Mehrdad Kazerani, Amir H. Hajimiragha et al. "Trends in microgrid control." *IEEE Transactions on Smart Grid* 5, no. 4 (2014): 1905–1919.

43. Wang, Huanhuan, Ashwin M. Khambadkone, and Xiaoxiao Yu. "Control of parallel connected power converters for low voltage microgrid—Part II: Dynamic electrothermal modeling." *IEEE Transactions on Power Electronics* 25, no. 12 (2010): 2971–2980.

44. Rokrok, Ebrahim, and Mohamad Esmail Hamedani Golshan. "Adaptive voltage droop scheme for voltage source converters in an islanded multibus microgrid." *IET Generation, Transmission & Distribution* 4, no. 5 (2010): 562–578.

45. Bidram, Ali, Vahidreza Nasirian, Ali Davoudi, and Frank L. Lewis. *Cooperative Synchronization in Distributed Microgrid Control.* Cham, Switzerland: Springer International Publishing, 2017.

46. He, Jinwei, and Yun Wei Li. "An enhanced microgrid load demand sharing strategy." *IEEE Transactions on Power Electronics* 27, no. 9 (2012): 3984–3995.

47. Lu, Xiaonan, Josep M. Guerrero, Kai Sun, and Juan C. Vasquez. "An improved droop control method for DC microgrids based on low bandwidth communication with DC bus voltage restoration and enhanced current sharing accuracy." *IEEE Transactions on Power Electronics* 29, no. 4 (2013): 1800–1812.

48. Zhong, Qing-Chang. "Harmonic droop controller to reduce the voltage harmonics of inverters." *IEEE Transactions on Industrial Electronics* 60, no. 3 (2012): 936–945.

49. Bidram, Ali, and Ali Davoudi. "Hierarchical structure of microgrids control system." *IEEE Transactions on Smart Grid* 3, no. 4 (2012): 1963–1976.

50. Li, Yun Wei, and Ching-Nan Kao. "An accurate power control strategy for power-electronics-interfaced distributed generation units operating in a low-voltage multibus microgrid." *IEEE Transactions on Power Electronics* 24, no. 12 (2009): 2977–2988.

51. Majumder, Ritwik, Arindam Ghosh, Gerard Ledwich, and Firuz Zare. "Angle droop versus frequency droop in a voltage source converter based autonomous microgrid." In *2009 IEEE Power & Energy Society General Meeting*, pp. 1–8. IEEE, Calgary, AB, Canada, 2009.

52. Palizban, Omid, and Kimmo Kauhaniemi. "Hierarchical control structure in microgrids with distributed generation: Island and grid-connected mode." *Renewable and Sustainable Energy Reviews* 44 (2015): 797–813.

53. Sun, Yao, Xiaochao Hou, Jian Yang, Hua Han, Mei Su, and Josep M. Guerrero. "New perspectives on droop control in AC microgrid." *IEEE Transactions on Industrial Electronics* 64, no. 7 (2017): 5741–5745.

54. Majumder, Ritwik, Balarko Chaudhuri, Arindam Ghosh, Rajat Majumder, Gerard Ledwich, and Firuz Zare. "Improvement of stability and load sharing in an autonomous microgrid using supplementary droop control loop." *IEEE Transactions on Power Systems* 25, no. 2 (2009): 796–808.

55. Yao, Wei, Min Chen, José Matas, Josep M. Guerrero, and Zhao-Ming Qian. "Design and analysis of the droop control method for parallel inverters considering the impact of the complex impedance on the power sharing." *IEEE Transactions on Industrial Electronics* 58, no. 2 (2010): 576–588.

56. Li, Yan, and Yun Wei Li. "Power management of inverter interfaced autonomous microgrid based on virtual frequency-voltage frame." *IEEE Transactions on Smart Grid* 2, no. 1 (2011): 30–40.

57. Li, Yan, and Yun Wei Li. "Decoupled power control for an inverter based low voltage microgrid in autonomous operation." In *2009 IEEE 6th International Power Electronics and Motion Control Conference*, pp. 2490–2496. IEEE, Wuhan, China, 2009.

58. Guerrero, Josep M., Juan C. Vasquez, José Matas, Luis García De Vicuña, and Miguel Castilla. "Hierarchical control of droop-controlled AC and DC microgrids—A general approach toward standardization." *IEEE Transactions on Industrial Electronics* 58, no. 1 (2010): 158–172.

59. Guerrero, Josep M., Lijun Hang, and Javier Uceda. "Control of distributed uninterruptible power supply systems." *IEEE Transactions on Industrial Electronics* 55, no. 8 (2008): 2845–2859.

60. He, Jinwei, Yun Wei Li, Josep M. Guerrero, Frede Blaabjerg, and Juan C. Vasquez. "An islanding microgrid power sharing approach using enhanced virtual impedance control scheme." *IEEE Transactions on Power Electronics* 28, no. 11 (2013): 5272–5282.

61. Han, Hua, Yao Liu, Yao Sun, Mei Su, and Josep M. Guerrero. "An improved droop control strategy for reactive power sharing in islanded microgrid." *IEEE Transactions on Power Electronics* 30, no. 6 (2014): 3133–3141.

62. Tuladhar, Anil, Hua Jin, Tom Unger, and Konrad Mauch. "Control of parallel inverters in distributed AC power systems with consideration of line impedance effect." *IEEE Transactions on Industry Applications* 36, no. 1 (2000): 131–138.

63. Kim, Jaehong, Josep M. Guerrero, Pedro Rodriguez, Remus Teodorescu, and Kwanghee Nam. "Mode adaptive droop control with virtual output impedances for an inverter-based flexible AC microgrid." *IEEE Transactions on power electronics* 26, no. 3 (2010): 689–701.

64. Sao, Charles K., and Peter W. Lehn. "Control and power management of converter fed microgrids." *IEEE Transactions on Power Systems* 23, no. 3 (2008): 1088–1098.

65. Zhong, Qing-Chang. "Robust droop controller for accurate proportional load sharing among inverters operated in parallel." *IEEE Transactions on Industrial Electronics* 60, no. 4 (2011): 1281–1290.

66. Sao, Charles K., and Peter W. Lehn. "Autonomous load sharing of voltage source converters." *IEEE Transactions on Power Delivery* 20, no. 2 (2005): 1009–1016.

67. Prodanovic, Milan, and Timothy C. Green. "High-quality power generation through distributed control of a power park microgrid." *IEEE Transactions on Industrial Electronics* 53, no. 5 (2006): 1471–1482.

68. Caldognetto, Tommaso, and Paolo Tenti. "Microgrids operation based on master–slave cooperative control." *IEEE Journal of Emerging and Selected Topics in Power Electronics* 2, no. 4 (2014): 1081–1088.

69. Andishgar, Mohammad Hadi, Eskandar Gholipour, and Rahmat-allah Hooshmand. "An overview of control approaches of inverter-based microgrids in islanding mode of operation." *Renewable and Sustainable Energy Reviews* 80 (2017): 1043–1060.

70. Tsikalakis, Antonis G., and Nikos D. Hatziargyriou. "Centralized control for optimizing microgrids operation." In *2011 IEEE Power and Energy Society General Meeting*, pp. 1–8. IEEE, Detroit, MI, USA, 2011.

71. Savaghebi, Mehdi, Alireza Jalilian, Juan C. Vasquez, and Josep M. Guerrero. "Secondary control for voltage quality enhancement in microgrids." *IEEE Transactions on Smart Grid* 3, no. 4 (2012): 1893–1902.

72. Mehrizi-Sani, Ali, and Reza Iravani. "Potential-function based control of a microgrid in islanded and grid-connected modes." *IEEE Transactions on Power Systems* 25, no. 4 (2010): 1883–1891.

73. Meng, Lexuan, Mehdi Savaghebi, Fabio Andrade, Juan C. Vasquez, Josep M. Guerrero, and Moisès Graells. "Microgrid central controller development and hierarchical control implementation in the intelligent microgrid lab of Aalborg University." In *2015 IEEE Applied Power Electronics Conference and Exposition (APEC)*, pp. 2585–2592. IEEE, Charlotte, NC, USA, 2015.

74. Babqi, Abdulrahman J., Zhehan Yi, and Amir H. Etemadi. "Centralized finite control set model predictive control for multiple distributed generator small-scale microgrids." In *2017 North American Power Symposium (NAPS)*, pp. 1–5. IEEE, Morgantown, WV, USA, 2017.

75. Tan, Kuan Tak, Xiao Yang Peng, Ping Lam So, Yun Chung Chu, and Michael Z. Q. Chen. "Centralized control for parallel operation of distributed generation inverters in microgrids." *IEEE Transactions on Smart Grid* 3, no. 4 (2012): 1977–1987.

76. Diaz, Nelson L., Adriana Carolina Luna, Juan C. Vasquez, and Josep M. Guerrero. "Centralized control architecture for coordination of distributed renewable generation and energy storage in islanded AC microgrids." *IEEE Transactions on Power Electronics* 32, no. 7 (2016): 5202–5213.

77. Karimi, Mazaher, Peter Wall, Hazlie Mokhlis, and Vladimir Terzija. "A new centralized adaptive underfrequency load shedding controller for microgrids based on a distribution state estimator." *IEEE Transactions on Power Delivery* 32, no. 1 (2016): 370–380.

78. Wang, Xiongfei, Frede Blaabjerg, Zhe Chen, and Josep M. Guerrero. "A centralized control architecture for harmonic voltage suppression in islanded microgrids." In *IECON 2011–37th Annual Conference of the IEEE Industrial Electronics Society*, pp. 3070–3075. IEEE, Melbourne, VIC, Australia, 2011.

79. Bidram, Ali, Ali Davoudi, Frank L. Lewis, and Shuzhi Sam Ge. "Distributed adaptive voltage control of inverter-based microgrids." *IEEE Transactions on Energy Conversion* 29, no. 4 (2014): 862–872.

80. Guo, Fanghong, Changyun Wen, Jianfeng Mao, and Yong-Duan Song. "Distributed secondary voltage and frequency restoration control of droop-controlled inverter-based microgrids." *IEEE Transactions on Industrial Electronics* 62, no. 7 (2014): 4355–4364.

81. Simpson-Porco, John W., Qobad Shafiee, Florian Dörfler, Juan C. Vasquez, Josep M. Guerrero, and Francesco Bullo. "Secondary frequency and voltage control of islanded microgrids via distributed averaging." *IEEE Transactions on Industrial Electronics* 62, no. 11 (2015): 7025–7038.

82. Kounev, Velin, David Tipper, Attila Altay Yavuz, Brandon M. Grainger, and Gregory F. Reed. "A secure communication architecture for distributed microgrid control." *IEEE Transactions on Smart Grid* 6, no. 5 (2015): 2484–2492.

83. Shafiee, Qobad, Josep M. Guerrero, and Juan C. Vasquez. "Distributed secondary control for islanded microgrids—A novel approach." *IEEE Transactions on Power Electronics* 29, no. 2 (2013): 1018–1031.

84. Shafiee, Qobad, Čedomir Stefanović, Tomislav Dragičević, Petar Popovski, Juan C. Vasquez, and Josep M. Guerrero. "Robust networked control scheme for distributed secondary control of islanded microgrids." *IEEE Transactions on Industrial Electronics* 61, no. 10 (2013): 5363–5374.

85. Ge, Xiaohua, and Qing-Long Han. "Distributed formation control of networked multi-agent systems using a dynamic event-triggered communication mechanism." *IEEE Transactions on Industrial Electronics* 64, no. 10 (2017): 8118–8127.

86. Heemels, W. P. Maurice Heemels, M. C. F. Tijis Donkers, and Andrew R. Teel. "Periodic event-triggered control for linear systems." *IEEE Transactions on Automatic Control* 58, no. 4 (2012): 847–861.

87. Zhang, Hao, Gang Feng, Huaicheng Yan, and Qijun Chen. "Observer-based output feedback event-triggered control for consensus of multi-agent systems." *IEEE Transactions on Industrial Electronics* 61, no. 9 (2013): 4885–4894.

88. Coelho, Ernane Antônio, Dan Wu, Josep M. Guerrero, Juan C. Vasquez, Tomislav Dragičević, C'edomir Stefanović, and Petar Popovski. "Small-signal analysis of the microgrid secondary control considering a communication time delay." *IEEE Transactions on Industrial Electronics* 63, no. 10 (2016): 6257–6269.

CHAPTER 4

Adaptive Distributed Secondary Control Schemes for Microgrid

4.1 INTRODUCTION AND LITERATURE REVIEW

In an MG, the energy generation is distributed in nature and can either be in the form of AC or DC [1]. To integrate these sources with the AC MG, power inverters are used; therefore, it is important to design a control strategy for the inverter that ensures (a) voltage regulation, (b) power balancing, (c) frequency regulation, (d) appropriate reactive power sharing, and (e) accurate active power sharing.

As discussed in Chapter 3, the control of an MG is generally practised in a hierarchical manner and is usually implemented in three different control architectures, namely: Decentralized, distributed, and centralized control architectures. Due to the high penetration of energy sources into the MG, the network dynamics become more complex; therefore, among these control architectures, a distributed structure is preferably used due to (a) PnP functionality, (b) robustness to single point failure, (c) low-bandwidth CN, (d) low cost of CN, and (e) easy and simple control structure [2]. Due to these advantageous features of distributed architecture, this chapter will mainly focus on the design of a distributed SC that regulates the frequency and voltage while ensuring accurate power sharing.

To eliminate the frequency and voltage mismatch between measured and nominal values caused by the primary controller, a distributed SC was

DOI: 10.1201/9781003594284-4

first proposed in Ref. [3]. In this work, a digraph or graph theory is used to achieve communication among the DGs. As it uses a low-bandwidth sparse CN, therefore, it only needed the information of the adjacent neighbouring DGs to accomplish the control tasks. The main disadvantages of this method are that it has a slow dynamic response and is unable to accurately regulate the voltage and ensure reactive power sharing at the same time. As the accurate voltage regulation leads to inappropriate reactive power sharing and vice versa. Hence, to cope with these challenges and improve the performance of the system, the authors in Refs. [4,5] proposed a PI-based distributed SC. Although the dynamic response of the system is improved but its lacks the ability to resolve the natural conflict between voltage regulation and reactive power sharing. Furthermore, in these works [3,5], high degradation in performance is observed in the presence of any uncertainty or disturbance.

To cope with these issues and resolve the conflict between voltage regulation and reactive power sharing, the authors in Ref. [6] proposed a Distributed Averaging Proportional Integral (DAPI) controller. In this method, the averaged values are taken as the nominal states to attain the information updating process between the DGs. Moreover, in case of small variation in load, this control shows efficient performance and regulates the frequency and voltage within acceptable limits while attaining appropriate power sharing. To further improve the performance of the controller, the authors in Ref. [7] extended this work and proposed a generalized averaging control scheme. This method successfully resolves a trade-off conflict between voltage regulation and reactive power sharing. The performance of this generalized DAPI control approach is validated in case of load variation and PnP scenarios. Although, as compared to the prior discussed methods, this method improves the system performance to a great extent; however, in case of any uncertainty (such as PnP functionality), high performance degradation can be observed.

To improve the controller's performance and to cope with the above-mentioned limitation, numerous authors have proposed different control strategies, such as the authors in Ref. [8] presented a two-layered distributed SC. In this method, a VSI is used in the first layer to regulate the frequencies and voltages of the DGs while in the second layer, a current source power inverter is utilized to ensure appropriate sharing of load among the DGs. A main disadvantage of this controller is its high dependency on system dynamics and DG parameters; moreover, the structural complexity of this controller is also high. A distributed SC for frequency regulation and active power sharing is proposed in Ref. [9]. The perfor-mance of the system in case of load variation is greatly enhanced by this controller. Moreover, in this controller, every agent measures the local active power and frequency, executes the computational procedures, and communicates with the neighbouring agents via CN, thus making it valu-able to perform PnP functionality. Besides these numerous advantages, this research lacks to consider the controller design, dynamics and per-formance for voltage regulation and reactive power sharing. To improve the transient performance of inverter-based MG, the authors in Ref. [10] proposed a DAPI control strategy. In this method, during transients, the resistive power losses are computed using H_2 norm and DAPI controller is used to minimize the losses. From the results, it can be concluded that the losses during the transients can be significantly reduced if a DAPI control-ler is tuned appropriately.

A pinning method based on distributed SC for voltage and frequency is proposed in Ref. [11]. In this technique, the computational burden in con-text with the exchange of information is significantly reduced as it uses less number of feedback controllers. In this method, one DG is chosen as a vir-tual leader; thus, it is required to pin only virtual leader DG, as all the other DGs will follow a leader DG. It shows high robustness in case of switch-ing CN, improves the stability of the system, and also facilitates the PnP operation. The authors in Ref. [12] proposed two distributed SC strategies based on iterative techniques. A first controller provides frequency regula-tion and active power sharing in a finite time. A second controller is imple-mented at a relatively slow time scale compared to the first controller and is responsible for voltage restoration and reactive power sharing. As these controllers work in a decoupled manner thus, it greatly reduces the over-shoots during transients. Similarly, another distributed architecture-based SC is presented in Ref. [13]. It attains a finite time convergence for fre-quency regulation and active power sharing, while for voltage regulation and reactive power sharing, it achieves an asymptotic convergence.

A consensus-based distributed MPC for voltage regulation is proposed in Ref. [14], in which every DG is permitted to perform its control tasks based on predictive actions. A two-layered finite-time consensus-based distributed SC is proposed in Ref. [15]. The mandatory SC actions such as voltage and frequency regulation, power-sharing, and balancing are achieved by this control technique. However, the mathematical model and stability of the finite-time controller are not taken into consideration. Similarly, another finite-time consensus-based distributed controller to improve the transient response and performance of the system in case of RES uncertainties is presented in Ref. [16]. This controller shows efficient performance in different case scenarios; however, during load variation conditions, high oscillation can be seen in output waveforms. Moreover, another drawback of this method is the dependency of voltage regulation on the model parameters. A distributed SC for frequency and voltage restoration proposed in Ref. [17] combines a fractional power integrator and sign function. This controller shows good performance in different operating conditions, such as load variation, switching CN topology, and also offers PnP functionality. However, besides these advantages, it suffers from chattering phenomena that may affect the stability of the system. The authors in Ref. [18] proposed a primal-dual gradient algorithm-based distributed SC for voltage regulation and reactive power sharing. In this control technique, the DGs' output voltage is considered as a control variable, whereas the output reactive power is considered as an objective function to successfully overcome the natural conflict between voltage regulation and reactive power sharing. This method shows a very effective performance, but the authors did not provide any explanation regarding frequency regulation and active power sharing.

4.2 CONTRIBUTIONS

From the above-discussed state-of-the-art literature, it can be concluded that the distributed SC must have the ability to (a) regulate the frequency and voltage according to its references while ensuring optimal power-sharing, (b) the frequency and voltage should be converged to their references synchronously and rapidly, and (c) the controller must have the ability to accommodate the uncertainties of DGs, physical network, and cyber network. To accomplish these goals of the SC, different researchers have designed numerous controllers and methodologies. However, to the best of the authors' knowledge, no work is still presented concerning the proposed adaptive control techniques that update the control law gain parameters of the SC to regulate the frequency and voltage of inverter-based MG while attaining appropriate power sharing. Moreover, distributed architecture eliminates the utilization of centralized supervisory control. Moreover, a new DG can be easily integrated into the existing system through a low-bandwidth CN. The main contributions of this chapter are:

- An MIT rule-based adaptive controller, BP-based API controller, and Lyapunov function-based adaptive controller are designed for the distributed SC of an MG.

- These adaptive-based control techniques update the SC gain parameters at every sampling time by continuously measuring the dynamic behaviour of the system, so that in case of uncertainty, high quality output waveforms and the stability of the system can be achieved.

- The proposed controllers do not require prior knowledge of the MG topology, loads, and impedances, thus enhancing the system performance; therefore, we can refer it as model-free.

4.3 PRELIMINARIES

The DGs are generally connected with an MG through power inverters; therefore, the control of the inverter is of great importance. Moreover, communication among the DGs and control actions are performed through CN. Therefore, in this sub-section, the details of the physical-cyber network, along with the design of the primary control level of the hierarchical structure, is discussed.

4.3.1 Cyber System

A CN in an MG system is considered as a bridge between physical system and its protection and control processes. A CN is an MG that can be modelled by using a diagraph that can be presented as $G = (N, \xi)$; where a set of nodes (i.e. DGs) is presented by N such that $N = \{1, \ldots \ldots n\}$; a set of edges (i.e. links) is denoted by ξ i.e., $\xi = \{e_1, \ldots, e_m\}$ such that $\xi = [N]^2$. An adjacent matrix of G can be presented as $A = R^{|N||N|}$ that consists of elements a_{cd}, where R presents a set of real numbers. If there is a link or edge between two neighbouring nodes i.e. nodes c and d then $a_{cd} = a_{dc} = 1$, else $a_{cd} = a_{dc} = 0$. The degree of c_{th}-node can be presented as $D_c = \sum\limits_{c,d \in |N|} a_{cd}$, where $c \neq d$. The Laplacian matrix (ℓ) of the diagraph can be given as $\ell = D - A$, where $D = \text{diag}(D_c) \in R^{|N||N|}$ represents the degree of the matrix. Moreover, the nodes are arranged in manner that if there is an edge or path between nodes c and d then the graph is said to be connected. Hence, if G is connected, then ℓ is a positive semi-definite matrix having an eigen-value of zero and have an eigen-vector 1_n, i.e. $\ell 1_n = 0_n$ [19].

4.3.2 Physical System

A generalized meshed MG system with n number of DGs is considered having an Admittance (Y_{cd}) between c_{th}-DG and d_{th}-DG and can be presented as $Y_{cd} = G_{cd} + jB_{cd} \in C$, in which $G_{cd} \in R$ and $B_{cd} \in R$ denotes the conductance and susceptance, respectively. In a scenario, when there is no link among c_{th}-DG and d_{th}-DG, then the admittance can be given as $Y_{cd} = 0$. On the contrary, when there is connection of c_{th}-DG with its neighbours then a set of neighbouring DGs can be presented as $N_c = \{d | d \in \mathbb{N}, c \neq d, Y_{cd} \neq 0\}$. Moreover, it is considered that MG system is strongly connected then for all the pairs $\{c,d\} \in \mathbb{N} \times \mathbb{N}, c \neq d$ in an MG there is a sequence of nodes from c and d such that any two consecutive connected nodes there exists a power line that shows the admittance [20].

Hence, on the basis of these considerations, the active and reactive power injected by c_{th}-DG can be presented as [21]:

$$P_c = E_c^2 G_{cc} - \sum_{d=1}^{n} E_c E_d |Y_{cd}| \cos(\delta_c - \delta_d - \phi_{cd}) \tag{4.1}$$

$$Q_c = -\left(\sum_{d=1}^{n} E_c E_d |Y_{cd}| \sin(\delta_c - \delta_d - \phi_{cd}) + E_c^2 B_{cc} \right) \tag{4.2}$$

where $|Y_{cd}| = \sqrt{G_{cd}^2 + B_{cd}^2}$ and the angle of admittance is denoted by ϕ_{cd}. Furthermore, it is considered that the transmission lines are purely inductive and lossless thus $G_{cd} = 0$, $Y_{cd} = jB_{cd}$, and $\phi_{cd} = \phi_{dc} = -(\pi/2)$. Hence, (4.1) and (4.2) on the basis of these considerations can be given as:

$$P_c = \sum_{d=1}^{n} E_c E_d |B_{cd}| \cos(\delta_c - \delta_d) \tag{4.3}$$

$$Q_c = E_c^2 \sum_{d=1}^{n} B_{cd} - \sum_{d=1}^{n} E_c E_d |B_{cd}| \cos(\delta_c - \delta_d) \tag{4.4}$$

Remark 4.1

The considerations made above are common and reasonable in power system analysis. To achieve purely inductive lines in a power system, the inductive effect should be more dominant than the resistive effect. A detailed justification and elaboration are presented in Refs. [20,22].

4.3.3 Primary Control Level

Generally, a primary control level consists of inverter output control and power-sharing control, which are discussed below in detail.

4.3.3.1 Inverter Output Control

Different controllers have been used to control the output current and voltage of the inverter in an MG; however, among them PR controllers are preferably used due to their high capability to track the sinusoidal references. At the resonant frequency, it offers an infinite gain, whereas at other frequencies, it provides almost no gain, as a result a zero steady-state error can be achieved easily [23]. Although a PR controller offers numerous advantages but the output current of GC inverter-based MG is not invincible for the harmonic content, which generally occurs due to power converters' non-linearities. Hence, to cope with the problem of harmonic contents, numerous advancements have been made in the PR controller (a detailed description of different PR controllers can be read in Ref. [24]).

The performance of the system is greatly improved by PR controllers present in state-of-the-art literature; however, a very least explanation about the tuning and design process of the controller is provided. Moreover, to reach the optimal values most researchers use trial and error method which is not very feasible. Some of the researchers explain the design and tuning procedures of their respective PR controllers, but they are implemented in a continuous time domain; however, most of the MG applications need implementation in the digital domain. Therefore, to cope with these issues, a PR controller is designed that has a simple tuning methodology and control architecture. As compared to traditional methodology that involves complex trigonometry, this research presents a methodological procedure to calculate the proportional (k_p) and resonant (k_r) gains along with the resonant path coefficients [25]. A simplified diagram of the primary control level that consists of a PR controller and a droop controller is shown in Figure 4.1.

In this controller, k_p is added with the resonant path that consists of k_r and a filter having a transfer function in z-domain is given as $H_r(z)$. A k_p can be calculated as:

$$k_p = \frac{(v)\sqrt{(v)}\omega_{res}\left(L_i + L_g\right) - \left(R_i + R_g\right)}{V_{DC}/2} \tag{4.5}$$

where $v = 2d_{am} + 1$ in which d_{am} presents a damping factor which value lies in between 0.9 and 1.0; L_i and L_g are the inverter and grid side inductances of LCL filter, respectively; R_i and R_g are the resistances of LCL filter at inverter and grid side, respectively; V_{DC} is the inverter's input voltage; $\omega_{res} = 2 \times \pi \times f_{res}$ is the resonant angular frequency (f_{res} is the resonant frequency) which must be equal to the nominal frequency. Moreover, the value of k_r and $H_r(z)$ can be calculates as:

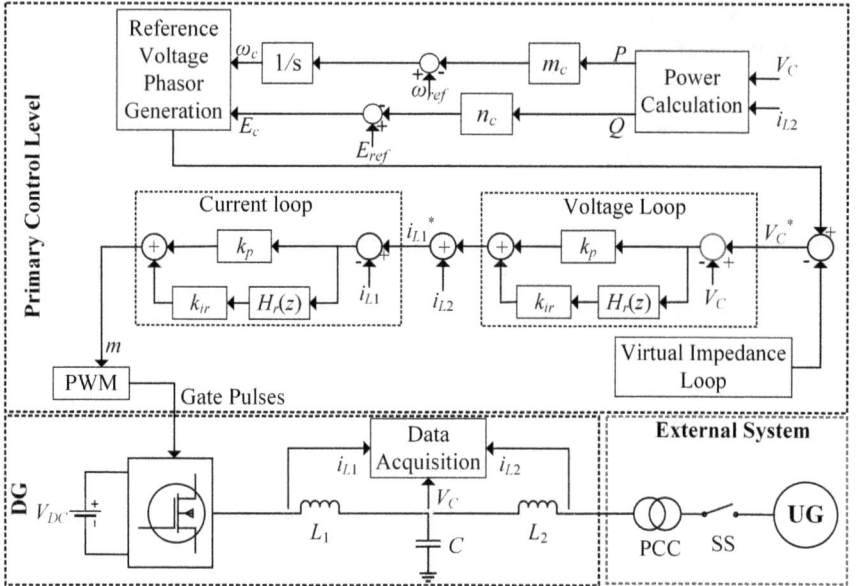

FIGURE 4.1 Schematic diagram of primary control level.

$$k_r = \frac{\omega_{res}^2(L_i + L_g)(v^2 - 1)}{V_{DC}} \tag{4.6}$$

$$H_r(z) = \frac{l_0 + l_1 z^{-1} + l_2 z^{-2}}{k_0 + k_1 z^{-1} + k_2 z^{-2}} \tag{4.7}$$

where l_0, l_1, l_2, k_0, k_1, and k_2 are the coefficients of $H_r(z)$ and can be calculated as:

$$k_0 = 1 \tag{4.8}$$

$$k_1 = -2e^{-0.5 B_{res} T_s} \cos\left(T_s\left(\sqrt{\omega_{res} - 0.25 B_{res}^2}\right)\right) \tag{4.9}$$

$$k_2 = e^{-B_{res} T_s} \tag{4.10}$$

$$l_0 = k_r T_s B_{res} \tag{4.11}$$

$$l_1 = \left[-B_{res} k_r e^{-0.5 T_s B_{res}} \cos\left(T_s \sqrt{\omega_{res}^2 - 0.25 B_{res}^2} - \mathbb{C} \right) \right] T_s \qquad (4.12)$$

$$l_2 = 0 \qquad (4.13)$$

where B_{res} is the resonant angular bandwidth and can be given as $B_{res} = 2 \times \pi \times B_s$; T_s denotes a sampling period; and $\mathbb{C}$ is a constant that can be given as:

$$\mathbb{C} = \sin\left(T_s \sqrt{\omega_{res}^2 - 0.25_{res}^2} \right) \frac{0.5 B_{res}^2 k_{res}}{\sqrt{\omega_r^2 - 0.25 B_r^2}} e^{-0.5 B_{res} T_s} \qquad (4.14)$$

To check the stability of the controller, it is simulated in a MATLAB® environment according to the parameters shown in Table 4.1. The simulation results of the resonant filter as presented in Figure 4.2a shows that only at 50 Hz all the elements are multiplied by 1, hence the gain turns to 0 dB while all the elements are attenuated and cannot pass the filter. Furthermore, at 50 Hz, due to the presence of poles in the transfer function of $H_r(z)$ a phase shift of 180° is observed that validates the performance and efficiency of $H_r(z)$. Likewise, when a proposed PR controller is simulated, the results of the magnitude and phase responses are shown in

TABLE 4.1   Parameters of PR Controller

Parameter	Value
DC-input voltage (V_{DC})	700 V
Damping factor (d_{am})	0.95
LCL inverter-side inductance (L_i)	1.74×10^{-4} H
LCL grid side inductance (L_g)	1.2×10^{-3} H
LCL inverter and grid side resistances (R_i, R_g)	0.01 ohm
LCL damping resistance (R_d)	20.5 ohm
LCL capacitance (C)	3.31×10^{-5} F
Resonant frequency (f_{res})	50 Hz
Resonant gain (k_r)	1
Resonant frequency bandwidth (B_{res})	1.5
Sampling period (T_s)	1×10^{-6} sec
Sampling frequency (f_a)	$1/1 \times 10^{-6}$ Hz
Switching frequency (f_{sw})	10×10^{-3} Hz

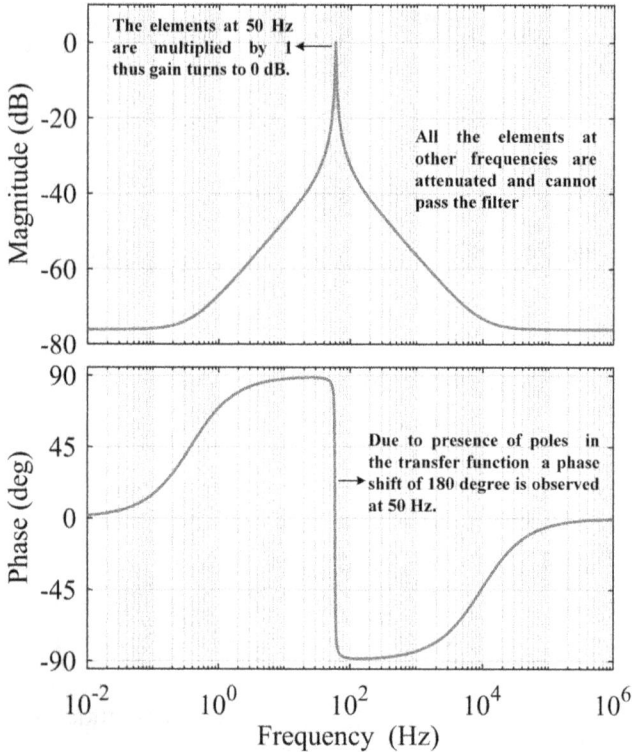

The elements at 50 Hz are multiplied by 1 thus gain turns to 0 dB.

All the elements at other frequencies are attenuated and cannot pass the filter

Due to presence of poles in the transfer function a phase shift of 180 degree is observed at 50 Hz.

FIGURE 4.2a Bode diagram of resonant filter.

Figure 4.2b. Similar, to a resonant filter, a highest amplification in magnitude is observed at 50 Hz which is equal to 47.4, and a phase shift of 180° occurs at 50 Hz, as can be seen in the phase response of the PR controller.

4.3.3.2 Example to Find the Values of Proportional gain, Resonant Gain, and the Components of Resonant Filter Using Table 4.1

The value of the proportional gain (k_p) according to (4.5) can be calculated if all the values of the term are known. From Table 4.1, the values of known parameters are as:

$$L_i = 1.74 \times 10^{-4} \, \text{H}$$

$$L_g = 1.2 \times 10^{-3} \, \text{H}$$

$$R_i = R_g = 0.01 \, \text{ohm}$$

$$V_{DC} = 700 \, \text{V}$$

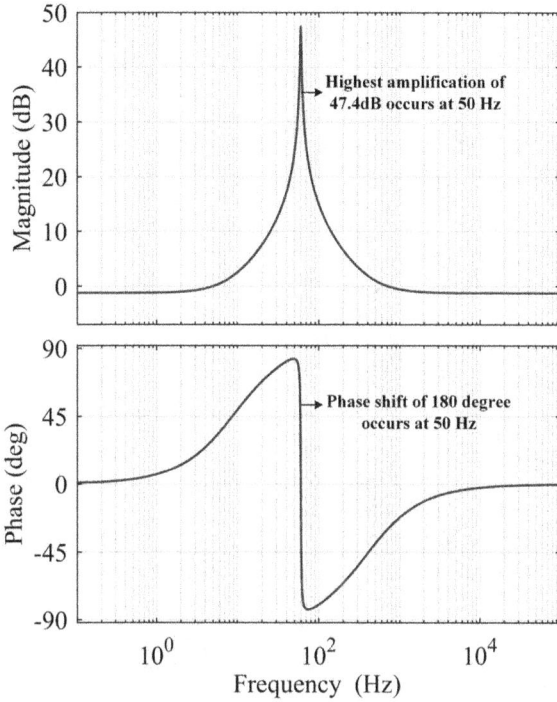

FIGURE 4.2b   Bode diagram of PR controller.

The other remaining unknown values of the parameters can be calculated as:

$$v = 2d_{am} + 1$$

$$v = 2(0.95) + 1 = 2.9$$

$$\omega_{res} = 2 \times \pi \times f_{res}$$

$$\omega_{res} = 2 \times 3.14 \times 50 = 314$$

Put all the values in (4.5) and solve it would yield us to:

$$k_p = \frac{(2.9)\sqrt{(2.9)}(314)\left(1.74 \times 10^{-4} + 1.2 \times 10^{-3}\right) - (0.01 + 0.1)}{700/2} = 7.075$$

Similarly, the value of resonant gain according to (4.6) can be calculated as:

$$k_r = \frac{(314)^2 (1.74 \times 10^{-4} + 1.2 \times 10^{-3})((2.9)^2 - 1)}{700} = 5.72$$

Finally, the components of resonant filter can be calculated according to (4.8–4.13) as:

$$k_0 = 1$$

$$k_1 = -2e^{-(0.5)(9.42)(1\times10^{-6})} \cos\left((1\times10^{-6})\left(\sqrt{314 - (.25)(9.42)^2}\right)\right) = -1.9999$$

$$k_2 = -e^{-(9.42)(1/10^{-6})} = 0.9999$$

$$l_0 = 1 \times (1 \times 10^{-6}) \times 9.42 = 9.4247 \times 10^{-6}$$

$$l_1 = \left[-(9.42)(1)e^{-(0.5)(1\times10^{-6})(9.42)} \cos\left(\begin{array}{c} (1\times10^{-6})\sqrt{(314)^2 - .25(9.42)^2} \\ -4.4413\times10^{-5} \end{array} \right) \right]$$

$$\times (1 \times 10^{-6}) = -9.4247 \times 10^{-6}$$

$$l_2 = 0$$

By using the above-discussed PR controller the exact values of every parameter can be achieved thus avoiding the trial and error methods which is mostly practiced in controllers presented in literature.

4.3.3.3 Power Sharing Control

To regulate the voltage and frequency while attaining accurate power-sharing $P - \omega$ and $Q - E$ droop control law for c_{th}-DG can be presented as [26]:

$$\omega_c = \omega_{ref} - m_c P_c \tag{4.15}$$

$$E_c = E_{ref} - n_c Q_c \tag{4.16}$$

where ω_c and E_c are the angular frequency and output voltage of c_{th}-DG, respectively $(c = 1,2,...n)$; E_{ref} and ω_{ref} are the nominal voltage and angular frequency, respectively, m_c and n_c are the P_c and Q_c droop coefficients, respectively, and can be given as:

$$m_c = \frac{(\omega_{max} - \omega_{min})}{P_{max}} \tag{4.17}$$

$$n_c = \frac{(E_{max} - E_{min})}{Q_{max}} \tag{4.18}$$

A droop controller can ensure power-sharing and stability of the system, but it causes a deviation of frequency and voltage from their nominal values. Hence, it automatically calls for the SC level of hierarchy that will be discussed in the upcoming sections of this chapter.

4.4 ADAPTIVE FREQUENCY REGULATION AND ACTIVE POWER SHARING CONTROLLER

A distributed controller for frequency regulation and active power sharing presented in Ref. [27] can be given as:

$$\omega_c = \omega_{ref} - m_c P_c + \varpi_c \tag{4.19}$$

$$\frac{d\varpi_c}{dt} = \mathbb{Z}_f \left[\sum_{d=1}^{n} a_{cd} (\varpi_d - \varpi_c) - (\omega_c - \omega_{ref}) \right] \tag{4.20}$$

where ϖ_c is the SC variable; $\mathbb{Z}_f$ is the gain of the SC; if there exists a cyber-link among c_{th}-DG and d_{th}-DG then $a_{cd} = 1$, however, on the contrary, when there is no link then $a_{cd} = 0$. A detail schematic diagram of the distributed control structure consists of physical system, primary controller, SC, and CN is presented in Figure 4.3.

If $\left[\sum_{d=1}^{n} a_{cd} (\varpi_d - \varpi_c) - (\omega_c - \omega_{ref}) \right] = 0$, then it can be said that the system is in steady-state and the following declarations becomes true: (a) In accordance with the nominal frequency all DGs adjust their frequencies, hence $\omega_c = \omega_d = \omega_{ref}$ and (b) same droop shift and an appropriate active power sharing is attained by all DGs, hence $\varpi_c = \varpi_d$.

From (4.20), it can be seen that the SC gain parameter $(\mathbb{Z}_f)$ is fixed and constant; therefore, in case of any uncertainty (PnP, load variation, etc.),

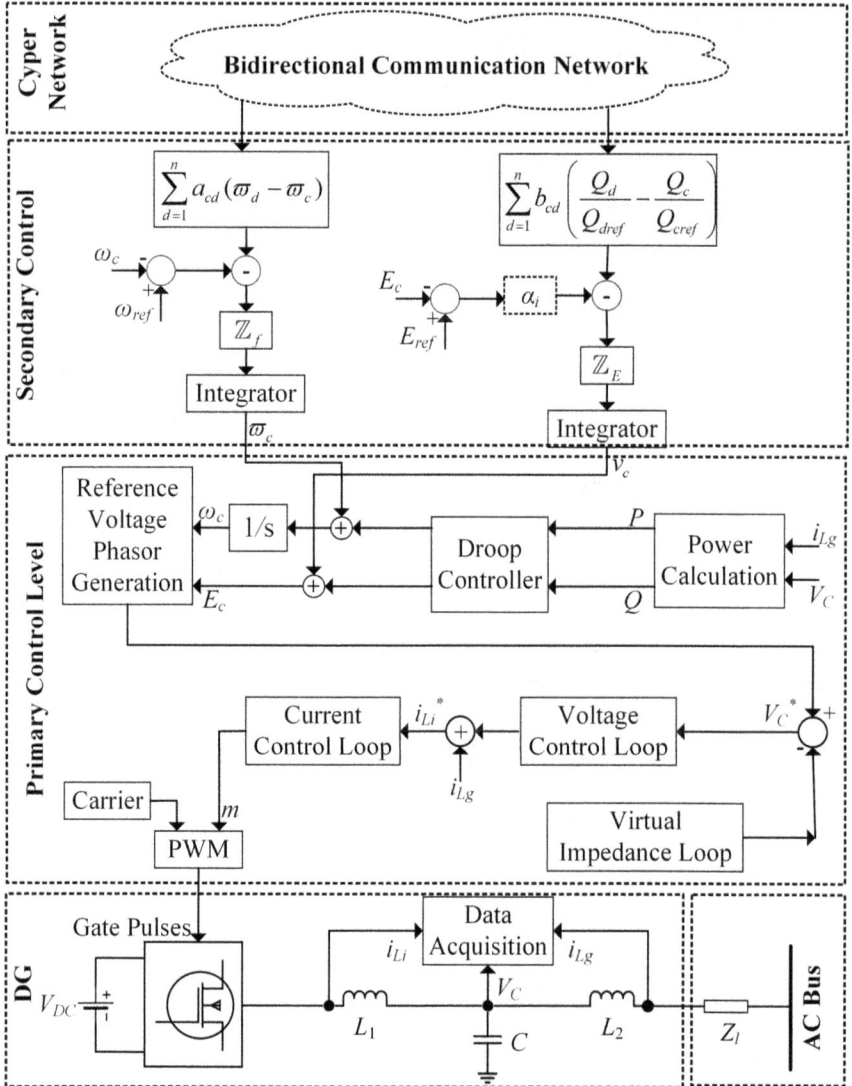

FIGURE 4.3 Generalized schematic of secondary controller.

its value is not updating which may affect the system's stability. Therefore, to cope with these challenges, in this chapter, three different adaptive control techniques are presented that update the value of $\mathbb{Z}_f$ at every sampling time to provide optimal controller performance.

4.4.1 BP-Based API Controller

A PI controller can be defined as:

$$\mathbb{Z}_f = k_{p-f}(t)e_f(t) + k_{i-f}(t)\int_0^t e_f(t)dt \tag{4.21}$$

where e_f is an error that can be given as $e_f = \omega - \omega_{ref}$. A BP technique that utilizes a gradient descent function to improve the dynamic response and minimize the error can be presented as [28]:

$$J = \frac{1}{2}e_f^2(t) \tag{4.22}$$

Hence, BP algorithm can be presented as:

$$k_{new} = -\beta\left(\frac{dJ}{dk_f}\right) \tag{4.23}$$

where β present a learning rate; $k_{new} \in \left[k_{p-f-new}, k_{i-f-new}\right]$ in which $k_{p-f-new}$ and $k_{i-f-new}$ presents the new proportional and integral gains. They can be calculated as:

$$k_{p-f-new} = -\beta\frac{dJ}{dk_{p-f}} = -\beta\left[\frac{\partial J}{\partial\omega}\frac{\partial\omega}{\partial\varpi}\frac{\partial\varpi}{\partial k_{p-f}}\right] = -\beta e_f^2(t)k_{p-f}(t) \tag{4.24}$$

$$k_{i-f-new} = -\beta\frac{dJ}{dk_{i-f}} = -\beta\left[\frac{\partial J}{\partial\omega}\frac{\partial\omega}{\partial\varpi}\frac{\partial\varpi}{\partial k_{i-f}}\right] = -\beta e_f^2(t)k_{i-f}(t) \tag{4.25}$$

Hence, the updated proportional and integral gain values can be given as:

$$k_{p-f}(t+1) = k_{p-f}(t) + k_{p-f-new}(t)$$
$$\Rightarrow k_{p-f}(t+1) = k_{p-f}(t)\left[1 - \beta e_f^2(t)\right] \tag{4.26}$$

$$k_{i-f}(t+1) = k_{i-f}(t) + k_{i-f-new}(t)$$
$$\Rightarrow k_{i-f}(t+1) = k_{i-f}(t)\left[1 - \beta e_f^2(t)\right] \tag{4.27}$$

Based on these updated values, the gains of the controller in (4.21) can be given as:

$$\overline{Z}_f = \left[\left\{ k_{p-f}(t)\left(1 - \beta e_f^2(t)\right) \right\} e_f(t) \right] + \left[\left\{ k_{i-f}(t)\left(1 - \beta e_f^2(t)\right) \right\} \int e_f(t) dt \right]$$

(4.28)

where $\overline{Z}_f$ is the updated gain of the SC, hence, a SC in (4.20) can be updated as:

$$\frac{d\varpi_c}{dt} = \left(\left[\left\{ k_{p-f}(t)\left(1 - \beta e_f^2(t)\right) \right\} e_f(t) \right] + \left[\left\{ k_{i-f}(t)\left(1 - \beta e_f^2(t)\right) \right\} \int e_f(t) dt \right] \right)$$

$$\left[\sum_{d=1}^{n} a_{cd}(\varpi_d - \varpi_c) - (\varpi_c - \varpi_{ref}) \right]$$

(4.29)

4.4.2 MIT Rule-Based Adaptive Controller

Before deriving an MIT rule-based adaptive controller an assumption is made as presented below:

Assumption 4.1

A control system presented in (4.19) can be described in updated and current form as:

$$\omega = \omega_{ref} - m_c P_c + y\varpi$$

(4.30)

$$\omega_z = \omega_{ref} - m_c P_c + y_z \varpi_z$$

(4.31)

where y denotes an updated gain and y_z presents a current gain value. Applying update control law, i.e. $\varpi = \zeta \varpi_z$ (ϖ and ϖ_z denotes the updated and previous control inputs and ζ presents an adjustment parameter) and put it in (4.30) would yield us to:

$$\omega = \omega_{ref} - m_c P_c + y\zeta \varpi_z$$

(4.32)

Let consider that updated error $\left(e_{f,z}\right)$ which varies with respect to time can be given as:

$$e_{f,z} = \omega - \omega_Z \tag{4.33}$$

Put (4.31) and (4.32) in (4.33) and take the time derivative on both sides, would give us:

$$\dot{e}_{f,z} = \dot{\omega}_{\text{ref}} - m_c \dot{P_c} + yz\dot{\omega}_Z - \dot{\omega}_{\text{ref}} + m_c \dot{P_c} - y\zeta\dot{\omega}_Z \tag{4.34}$$

Solving (4.34) would give us:

$$\dot{e}_{f,z} = yz\dot{\omega}_Z - y\zeta\dot{\omega}_Z \tag{4.35}$$

At a steady-state $e_{f,z} = 0$, hence, $\dot{e}_{f,z} = 0$. Therefore,

$$\zeta = \frac{y_Z}{y} \tag{4.36}$$

Consider a control system given in (4.20) with Assumption 4.1, an MIT rule can be defined as [29]:

$$J(\zeta) = \frac{1}{2}\left(e_{f,z}\right)^2 = \frac{1}{2}\left(\omega - \omega_Z\right)^2 \tag{4.37}$$

In (4.37), ζ is adjusted to minimize the loss function so it is logical to change the parameters in the negative direction of gradient J, thus we get:

$$\frac{d\zeta}{dt} = -\chi\frac{\partial J}{\partial \zeta} \tag{4.38}$$

where χ is the learning rate, thus:

$$\frac{dJ}{d\zeta} = \frac{\partial J}{\partial \omega}\cdot\frac{\partial \omega}{\partial \varpi}\cdot\frac{\partial \varpi}{\partial \zeta} = \zeta\left(e_{f,z}\right)^2 \tag{4.39}$$

Put (4.39) in (4.38), we get:

$$\frac{d\zeta}{dt} = -\chi\zeta\left(e_{f,z}\right)^2 \tag{4.40}$$

This equation, i.e. (4.40), is used to update the value of $\mathbb{Z}_f$ in (4.20). Thus, the adaptive-based distributed SC can be given as:

$$\frac{d\varpi_c}{dt} = \left(-\chi\zeta\left(e_{f,z}\right)^2\right)\left[\sum_{d=1}^{n} a_{cd}(\varpi_d - \varpi_c) - (\varpi_c - \varpi_{ref})\right] \tag{4.41}$$

4.4.3 Lyapunov Function-Based Adaptive Controller

Consider a control system given in (4.20) with Assumption 4.1, a Lyapunov function can be given as:

$$V(e_{f,z},\zeta) = \frac{1}{2}\kappa e_{f,z}^2 + \frac{1}{2}y(\zeta - \zeta_z)^2 \tag{4.42}$$

where κ presents a learning rate. Taking a derivative of (4.42) on both sides would yield us to:

$$\dot{V} = \kappa e_{f,z}\frac{de_{f,z}}{dt} + y(\zeta - \zeta_z)\frac{d\zeta}{dt} \tag{4.43}$$

Let consider $\zeta_z = y_z/y$. Put $\zeta_z = y_z/y$ and $\varpi = \zeta\varpi_z$ in (4.43), we get:

$$\dot{V} = \kappa e_{f,z}y\zeta\frac{d\varpi_z}{dt} - \kappa e_{f,z}y_z\frac{d\varpi_z}{dt} + (y\zeta - y_z)\frac{d\zeta}{dt} \tag{4.44}$$

Solving (4.44) would give us:

$$\dot{V} = (y\zeta - y_z)\left(\kappa e_{f,z}\frac{d\varpi_z}{dt} + \frac{d\zeta}{dt}\right) \tag{4.45}$$

In (4.45), if $d\zeta/dt = -\kappa e_{f,z}\left(d\varpi_z/dt\right)$, then $\dot{V} \to 0$ that signifies the convergence. Thus, we have:

$$\zeta = -\int\left(\kappa e_{f,z}\left(\frac{d\varpi_z}{dt}\right)\right) \tag{4.46}$$

As $\varpi = \zeta \varpi_z$, thus (4.46) can be written as:

$$\varpi = -\varpi_z \int\left(\kappa e_{f,z}\left(\frac{d\varpi_z}{dt}\right)\right) \tag{4.47}$$

An Expression in (4.47) is used to update the $\mathbb{Z}_f$ in (4.20). Thus (4.20) can be expressed as:

$$\frac{d\varpi_c}{dt} = \left[-\varpi_z \int\left(\kappa e_{f,z}\left(\frac{d\varpi_z}{dt}\right)\right)\right]\left[\sum_{d=1}^{n} a_{cd}(\varpi_d - \varpi_c) - (\varpi_c - \varpi_{ref})\right] \tag{4.48}$$

Remark 4.2

The above-discussed adaptive method is asymptotically stable conferring to Lyapunov theorem and provides enough conditions for convergence and stability of the system [30].

The adaptive controllers discussed above ensure that every DG operates at its nominal frequencies while maintaining accurate sharing of load among the DGs even, if some uncertainty such as PnP and load variation occurs.

4.5 ADAPTIVE VOLTAGE REGULATION AND REACTIVE POWER SHARING CONTROLLER

A distributed controller for voltage regulation and reactive power sharing proposed in Ref. [27] can be given as:

$$E_c = E_{ref} - n_c Q_c + v_c \tag{4.49}$$

$$\dot{v}_c = \mathbb{Z}_E\left[\sum_{d=1}^{n} b_{cd}\left(\frac{Q_d}{Q_{d\,ref}} - \frac{Q_c}{Q_{c\,ref}}\right) - \alpha_c(E_c - E_{ref})\right] \tag{4.50}$$

where v_c is the SC variable; $\mathbb{Z}_E$ and α_c are the designed gains of the SC; if there exists a cyber-link among c_{th}-DG and d_{th}-DG then $b_{cd} = 1$, on the contrary, when there is no link then $b_{cd} = 0$, moreover, to avoid extra CN it is assumed that $B = A \Rightarrow \{b_{ij}\} = \{a_{ij}\}$.

If a system is in steady-state, then the following statements become true: (a) In accordance with the nominal voltage all DGs adjust their voltages hence it can be written as $E_c = E_d = E_{ref}$ and (b) a proportionate reactive power sharing has been attained, it means that $\left(Q_c/Q_{c\,ref}\right) = \left(Q_d/Q_{d\,ref}\right)$.

From (4.50), it can be seen the SC gain parameter $\left(\mathbb{Z}_E\right)$ is constant, therefore, in case of any uncertainty its value is not updating that may affect the system's stability. Therefore, the adaptive controllers are applied, that update the value of $\mathbb{Z}_E$ to provide an optimal controller performance.

4.5.1 BP-Based API Controller

A PI controller can be defined as:

$$\mathbb{Z}_E = k_{p-E}(t)e_E(t) + k_{i-E}(t)\int_0^t e_E(t)dt \qquad (4.51)$$

where e_E is an error that can be given as $e_E = E - E_{ref}$. A BP technique that utilizes a gradient descent function to improve the dynamic response and minimize the error can be presented as:

$$J = \frac{1}{2}e_E^2(t) \qquad (4.52)$$

Hence, to update $\mathbb{Z}_E$ a BP algorithm can be presented as:

$$k_{new} = -\varepsilon\left(\frac{dJ}{dk_E}\right) \qquad (4.53)$$

where ε present a learning rate; $k_{new} \in \left[k_{p-E-new}, k_{i-E-new}\right]$ in which $k_{p-E-new}$ and $k_{i-E-new}$ presents the new proportional and integral gains. They can be calculated as:

$$k_{p-E-new} = -\varepsilon\frac{dJ}{dk_{p-E}} = -\varepsilon\left[\frac{\partial J}{\partial E}\frac{\partial E}{\partial v}\frac{\partial v}{\partial k_{p-E}}\right] = -\varepsilon e e_E^2(t)k_{p-E}(t) \quad (4.54)$$

$$k_{i-E-new} = -\varepsilon\frac{dJ}{dk_{i-E}} = -\varepsilon\left[\frac{\partial J}{\partial E}\frac{\partial E}{\partial v}\frac{\partial v}{\partial k_{i-E}}\right] = -\varepsilon e e_E^2(t)k_{i-E}(t) \quad (4.55)$$

Hence, the updated proportional and integral gain values can be given as:

$$k_{p-E}(t+1)=k_{p-E}(t)+k_{p-E-\text{new}}(t)=k_{p-E}(t)\left[1-\varepsilon e_E^2(t)\right] \quad (4.56)$$

$$k_{i-E}(t+1)=k_{i-E}(t)+k_{i-E-\text{new}}(t)=k_{i-E}(t)\left[1-\varepsilon e_E^2(t)\right] \quad (4.57)$$

Based on these updated values, the gains of the controller in (4.51) can be given as:

$$\bar{Z}_E=\left[\left\{k_{p-E}(t)\left(1-\varepsilon e_E^2(t)\right)\right\}e_E(t)\right]+\left[\left\{k_{i-E}(t)\left(1-\varepsilon e_E^2(t)\right)\right\}\int e_E(t)dt\right] \quad (4.58)$$

where $\bar{Z}_E$ is the updated gain of the SC, hence, a SC in (4.50) can be updated as:

$$\dot{v}_c=\left[\left\{k_{p-E}(t)\left(1-\varepsilon e_E^2(t)\right)\right\}e_E(t)\right]+\left[\left\{k_{i-E}(t)\left(1-\varepsilon e_E^2(t)\right)\right\}\int e_E(t)dt\right]$$

$$\times\left[\sum_{d=1}^n b_{cd}\left(\frac{Q_d}{Q_{d\,\text{ref}}}-\frac{Q_c}{Q_{c\,\text{ref}}}\right)-\alpha_c(E_c-E_{\text{ref}})\right] \quad (4.59)$$

4.5.2 MIT Rule-Based Adaptive Controller

An assumption should be made before deriving an MIT rule for updating $\mathbb{Z}_E$. It is given below as:

Assumption 4.2

A control system presented in (4.49) can be described in updated and current form as:

$$E=E_{\text{ref}}-n_cQ_c+xv \quad (4.60)$$

$$E_Z=E_{\text{ref}}-n_cQ_c+x_Zv_Z \quad (4.61)$$

where x denotes an updated gain and x_Z presents a current gain value. Applying update control law, i.e. $v=\xi v_Z$ (v and v_Z denotes the updated and previous control inputs and ξ presents an adjustment parameter) and put it in (4.60) would yield us to:

$$E = E_{ref} - n_c Q_c + x\xi v_Z \qquad (4.62)$$

Let consider that updated error $\left(e_{E,Z}\right)$ which varies with respect to time can be given as:

$$e_{E,Z} = E - E_Z \qquad (4.63)$$

Put (4.61) and (4.62) in (4.63) and take the time derivative on both sides, would give us:

$$\dot{e}_{E,Z} = \dot{E}_{ref} - n_c \dot{Q}_c + x_Z \dot{v}_Z - \dot{E}_{ref} + n_c \dot{Q}_c - x\xi \dot{v}_Z \qquad (4.64)$$

Solving (4.64) would give us:

$$\dot{e}_{E,Z} = x_Z \dot{v}_Z - x\xi \dot{v}_Z \qquad (4.65)$$

At a steady-state $e_{E,Z} = 0$, hence, $\dot{e}_{E,Z} = 0$. Therefore,

$$\xi = \frac{x_Z}{x} \qquad (4.66)$$

Consider a control system given in (4.50) with Assumption 4.2, an MIT rule can be defined as:

$$J(\xi) = \frac{1}{2}\left(e_{E,Z}\right)^2 = \frac{1}{2}\left(E - E_Z\right)^2 \qquad (4.67)$$

In (4.67), ξ is adjusted to minimize the loss function so it is logical to change the parameters in the negative direction of gradient J, thus we get:

$$\frac{d\xi}{dt} = -\phi \frac{\partial J}{\partial \xi} \qquad (4.68)$$

where ϕ is the learning rate, hence:

$$\frac{dJ}{d\xi} = \frac{\partial J}{\partial E} \cdot \frac{\partial E}{\partial v} \cdot \frac{\partial v}{\partial \xi} = \xi\left(e_{E,Z}\right)^2 \qquad (4.69)$$

Put (4.69) in (4.68), and (4.58) in (4.70) we get:

$$\frac{d\xi}{dt} = -\phi\xi\left(e_{E,Z}\right)^2 \tag{4.70}$$

This equation (4.70) is used to update the value of $\mathbb{Z}_E$ in (4.50). Thus the adaptive-based distributed SC can be given as:

$$\frac{dv_c}{dt} = \left(-\phi\xi\left(e_{E,Z}\right)^2\right)\left[\sum_{d=1}^{n} b_{cd}\left(\frac{Q_d}{Q_{d\,ref}} - \frac{Q_c}{Q_{c\,ref}}\right) - \alpha_c\left(E_c - E_{ref}\right)\right] \tag{4.71}$$

4.5.3 Lyapunov Function-Based Adaptive Controller

Consider a control system given in (4.50) with Assumption 4.2, a Lyapunov function can be given as:

$$V(e_{E,Z},\xi) = \frac{1}{2}\varphi e_{E,Z}^2 + \frac{1}{2}x(\xi - \xi_Z)^2 \tag{4.72}$$

where φ presents a learning rate. Taking a derivative of (4.72) on both sides would yield us to:

$$\dot{V} = \varphi e_{E,Z}\frac{de_{E,Z}}{dt} + x(\xi - \xi_Z)\frac{d\xi}{dt} \tag{4.73}$$

Let's consider $\xi_Z = x_Z/x$ and put it along with $v = \xi v_Z$ in (4.73), we get:

$$\dot{V} = \varphi e_{E,Z}x\xi\frac{dv_Z}{dt} - \varphi e_{E,Z}x_Z\frac{dv_Z}{dt} + (x\xi - x_Z)\frac{d\xi}{dt} \tag{4.74}$$

Solving (4.74) would give us:

$$\dot{V} = \left(x\xi - x_Z\right)\left(\varphi e_{E,Z}\frac{dv_Z}{dt} + \frac{d\xi}{dt}\right) \tag{4.75}$$

In (4.75), if $d\xi/dt = -\varphi e_{E,Z}\left(dv_Z/dt\right)$, then $\dot{V} \to 0$ that signifies a convergence. Thus, we have:

$$\xi = -\int\left(\varphi e_{E,Z}\left(\frac{dv_Z}{dt}\right)\right) \tag{4.76}$$

As $v = \xi v_Z$, thus (4.76) can be written as:

$$v = -v_Z \int \left(\varphi e_{E,Z} \left(\frac{dv_Z}{dt} \right) \right) \tag{4.77}$$

An Expression in (4.77) is used to update the $\mathbb{Z}_E$ in (4.50) and can be expressed as:

$$\frac{dv_c}{dt} = \left[-v_Z \int \left(\varphi e_{E,Z} \left(\frac{dv_Z}{dt} \right) \right) \right] \left[\sum_{d=1}^{n} b_{cd} \left(\frac{Q_d}{Q_{d\,\text{ref}}} - \frac{Q_c}{Q_{c\,\text{ref}}} \right) - \alpha_c \left(E_c - E_{\text{ref}} \right) \right] \tag{4.78}$$

These adaptive controllers ensure that every DG operates at their nominal voltage while maintaining accurate reactive power sharing even if some uncertainty such as PnP, load variation, etc. occurs.

4.6 SIMULATION RESULTS AND DISCUSSION

To validate the efficacy and performance of the adaptive distributed SCs, simulations are conducted in MATLAB/SimPower environment. An MG system that is considered in this work consists of 4 radial buses, 4-DGs (DG_1–DG_4), 4 local loads (L_1–L_4), and 1 public load (L_0) as presented in Figure 4.4. The values of different parameters and rating used in this simulation study are presented in Table 4.2. Furthermore, as this study is based on distributed architecture, therefore, every DG is linked with its neighbouring DGs through CN which adjacent matrix $A = B = [a_{cd}] = [b_{cd}]$ can be presented as:

$$A = B = \begin{pmatrix} 0 & 1 & 0 & 1 \\ 1 & 0 & 1 & 0 \\ 0 & 1 & 0 & 1 \\ 1 & 0 & 1 & 0 \end{pmatrix} \tag{4.79}$$

To validate the proposed controllers' performance two case studies (PnP functionality and load variation condition) are conducted that are explained below in detail [31,32].

4.6.1 Plug-and-Play Functionality

As the energy generation from RESs is stochastic and intermittent in nature; therefore, an MG must be able to allow a DG to be attached or

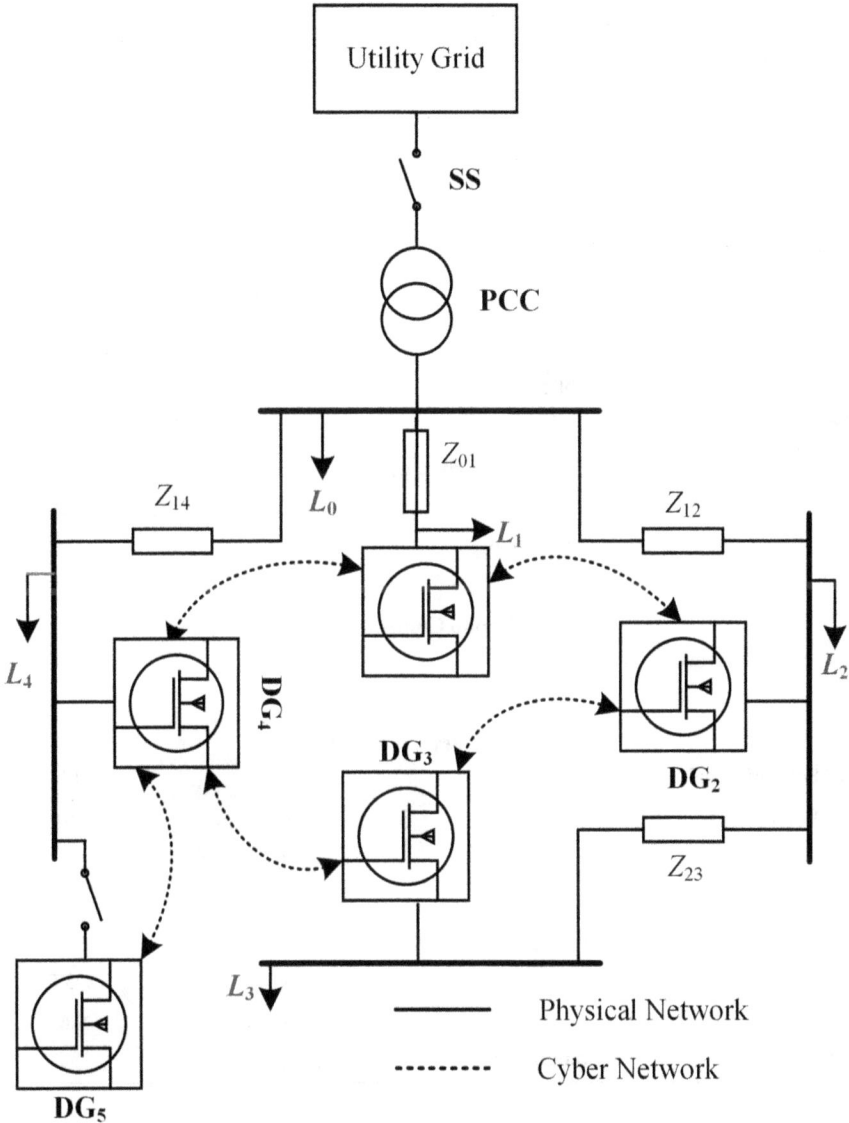

FIGURE 4.4 MG system under consideration.

detached at any instant without redesigning its existing structure. Hence, it is very important for the stable and smooth operation of an MG that the controller must be able to regulate the frequency and voltage along with accurate sharing of load among the DGs if a system is under such conditions.

TABLE 4.2   Parameters Used in this Study

Parameter		Symbol	Value
Electrical			
Frequency	Switching	f_{sw}	10×10^{-3} Hz
	Nominal	f	50 Hz
	Sampling	f_a	$1/1 \times 10^{-6}$ Hz
Nominal Voltage		E	310 V
Line Impedance		Z_{01}, Z_{14}	$0.8\,\Omega + 3$ mH
		Z_{12}	$1.6\,\Omega + 6$ mH
		Z_{23}	$0.9\,\Omega + 4$ mH
	Public	L_0	2 kW + 3 kVar
Loads	Local	$L_1 = L_2 = L_3 = L_4$	1 kW + 1 kVar
LCL Filter			
Capacitance		C_f	3.31×10^{-5} F
Inductance	Inverter side	L_i	1.74×10^{-4} H
	Grid side	L_g	1.2×10^{-3} H
Droop Controller			
Droop coefficients	P-ω	m_c	10^{-5} rad/(W s)
	Q-E	n_c	10^{-3} V/Var
Secondary Controller			
SC Frequency gain		$\mathbb{Z}_f$	0.1
SC Voltage gain		$\mathbb{Z}_E$	0.01
		β	2.4
Learning rates	BP-based API	ε	3.6
	MIT rule	χ	284.4
		ϕ	293.5
		κ	9.6
	Lyapunov Function	φ	15.2

4.6.1.1 Distributed SC Presented in Ref. [27]

It is considered that from $0 \leq t \leq 1$, an MG performs its normal operation; however, at $t=1$ sec, a DG_5 is plugged-in into the system and plugged-out of the system at $t=2$ sec (cyber-physical networks are also connected and disconnected), as can be seen in Figure 4.4. Initially, the performance of the conventional distributed SC presented in Ref. [27] is applied to check its performance during PnP operation. When DG_5 is plugged-in into a system, it synchronizes with existing DGs and generates a new power consensus, as can be seen in Figure 4.5. At $t=1$ sec, i.e., during plug-in operation, deviations in the output frequency waveforms of the DGs are observed, but the controller restores them and comes to its steady state within 0.45 sec with a deviation of 0.4 Hz as presented in Figure 4.5a. Likewise, in case of voltage regulation, a deviation of 12 V from its nominal value is observed, but the controller restores these deviations within 0.4 sec, as shown in Figure 4.5b. It should be noted that in case of voltage regulation during PnP operation, an SC with constant gain parameters is unable to perform efficiently, and an arbitrary consensus below or above from its reference is generated, which is not feasible. Moreover, during the plug-in operation, a new power consensus is generated and the load is proportionally shared among all the DGs, as can be seen in Figure 4.5c and d. In case of plug-out operation (when DG_5 is plug-out at $t=1$ sec), the same performance of the controller is observed as presented in Figure 4.5.

From the simulation results presented in Figure 4.5, it can be observed that the DAPI controller with constant gain parameters lacks the ability to precisely track the reference and is unable to accurately handle any uncertainty. Moreover, a low dynamic response is also observed by the DAPI controller. All these points motivate the authors to present the adaptive controllers that update the parameters of the SC in such scenarios.

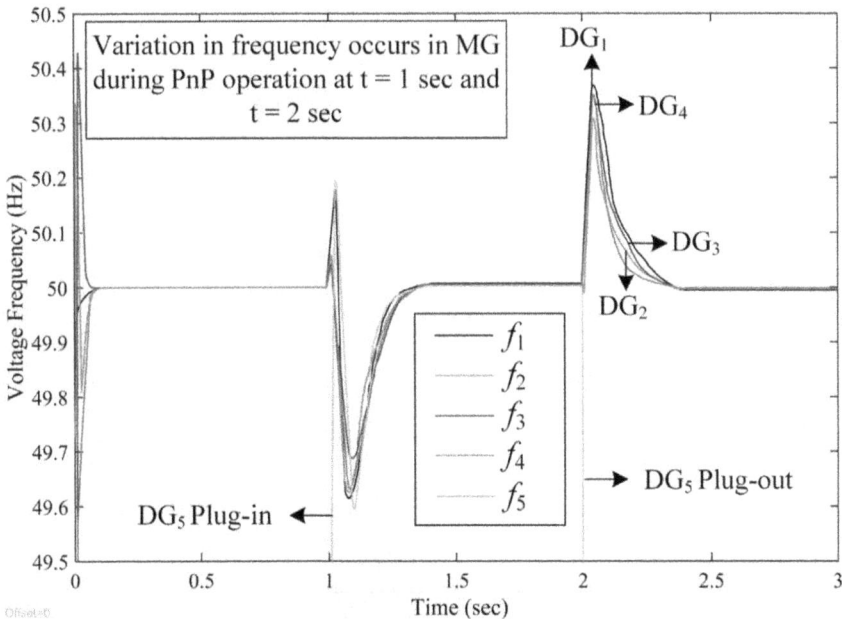

FIGURE 4.5a Frequency waveforms during PnP functionality by using DAPI controller presented in Ref. [27].

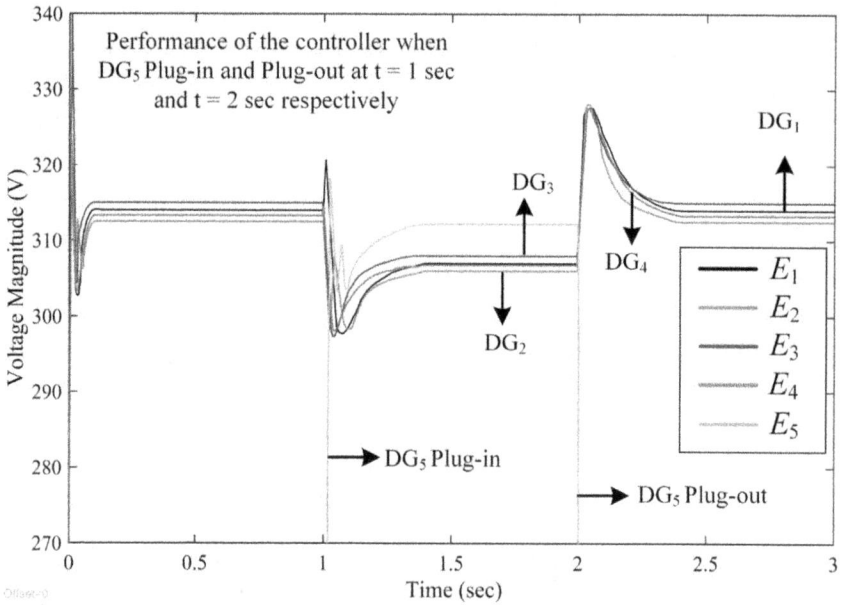

FIGURE 4.5b Voltage waveforms during PnP functionality by using DAPI controller presented in Ref. [27].

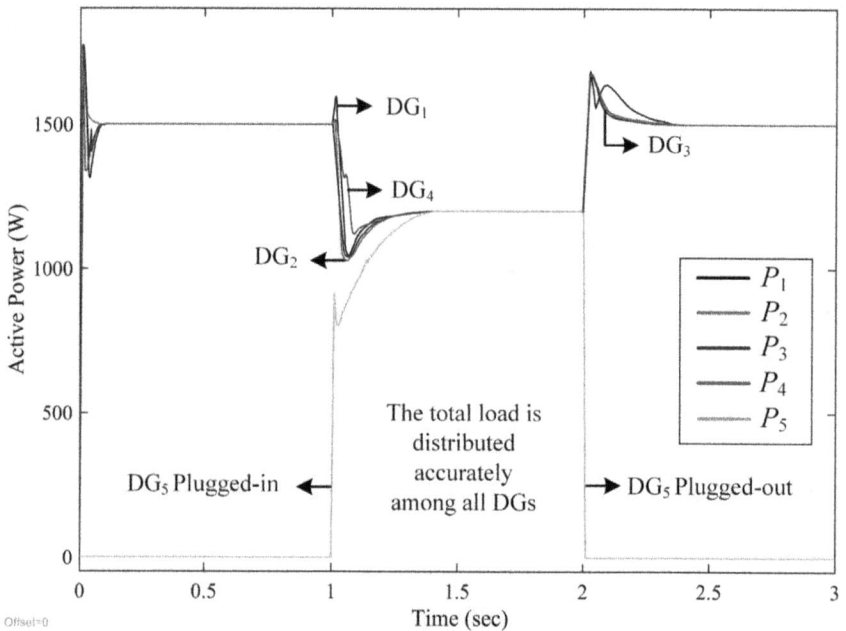

FIGURE 4.5c Active power waveforms during PnP functionality by using DAPI controller presented in Ref. [27].

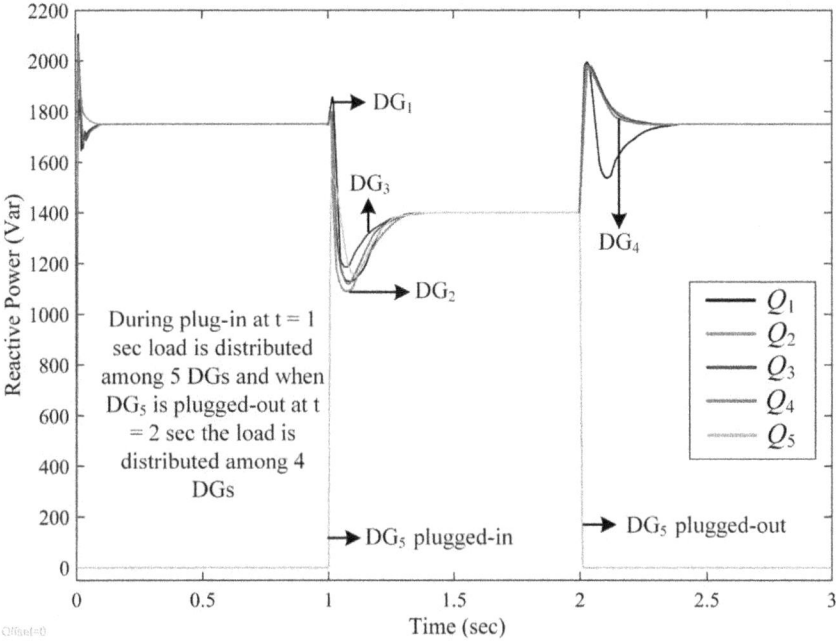

FIGURE 4.5d Reactive power waveforms during PnP functionality by using DAPI controller presented in Ref. [27].

4.6.1.2 BP-Based API Control Scheme

To improve the dynamic response and capability of the DAPI to perform efficiently in PnP operations, a BP-based API controller is proposed. To show the effectiveness of this controller, simulation studies are conducted as shown in Figure 4.6. The output frequency and voltage waveforms of DGs during the plug-in and plug-out phenomena are presented in Figure 4.6a and b. From the waveforms, it can be observed that during PnP operation, a deviation in waveforms from their references is observed, but the proposed controller restores these deviations with a fast dynamic response, and within 0.34 sec, they come to their steady-state, having a maximum deviation of 0.24 Hz in frequency and 9 V in voltage. At the same time, during PnP operation, a new power consensus is generated at $t=1$ sec and $t=2$ sec and the loads are accurately distributed among all the DGs as can be seen in Figure 4.6c and d. From Figure 4.6, it can be observed that although the performance of the controller as compared to [27] is considerably improved but still the controller is not able to accurately track the nominal voltage.

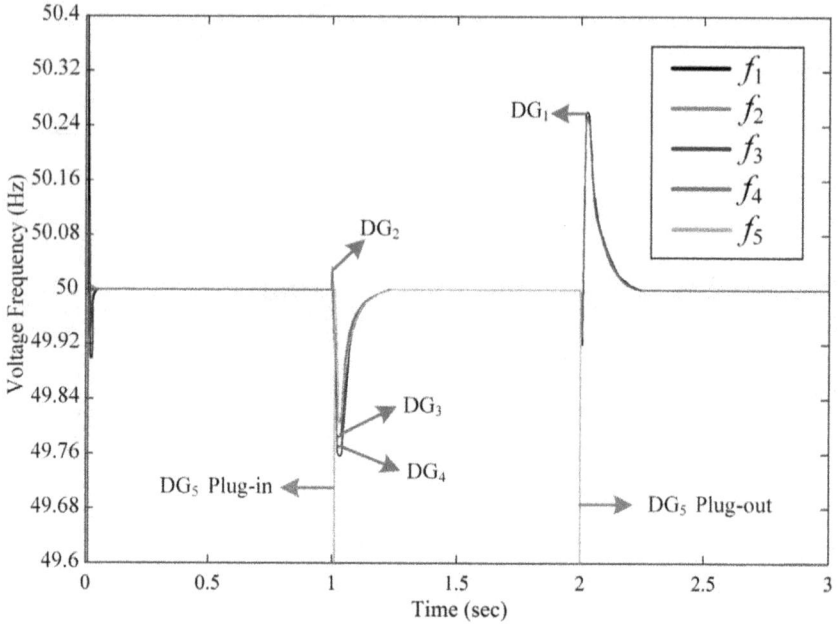

FIGURE 4.6a   Frequency waveforms during PnP functionality by using BP-based API controller.

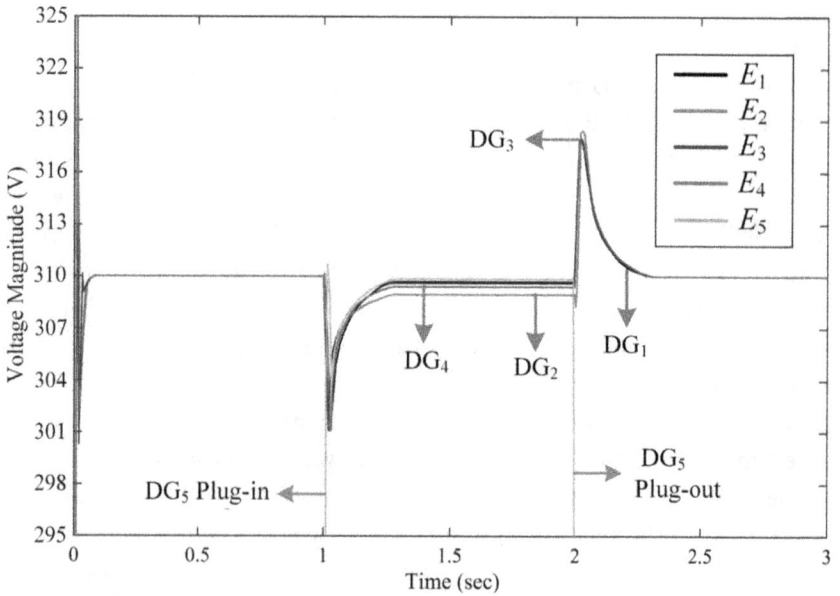

FIGURE 4.6b   Voltage waveforms during PnP functionality by using BP-based API controller.

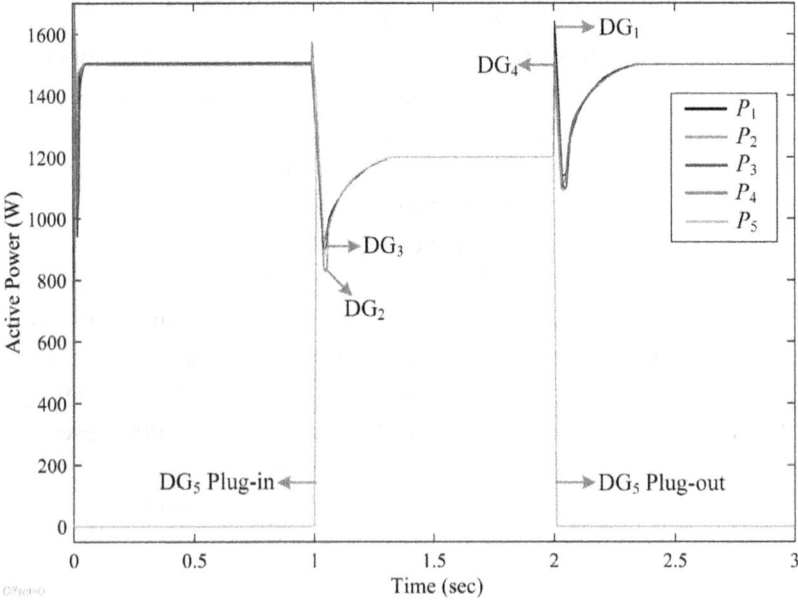

FIGURE 4.6c Active power waveforms during PnP functionality by using BP-based API controller.

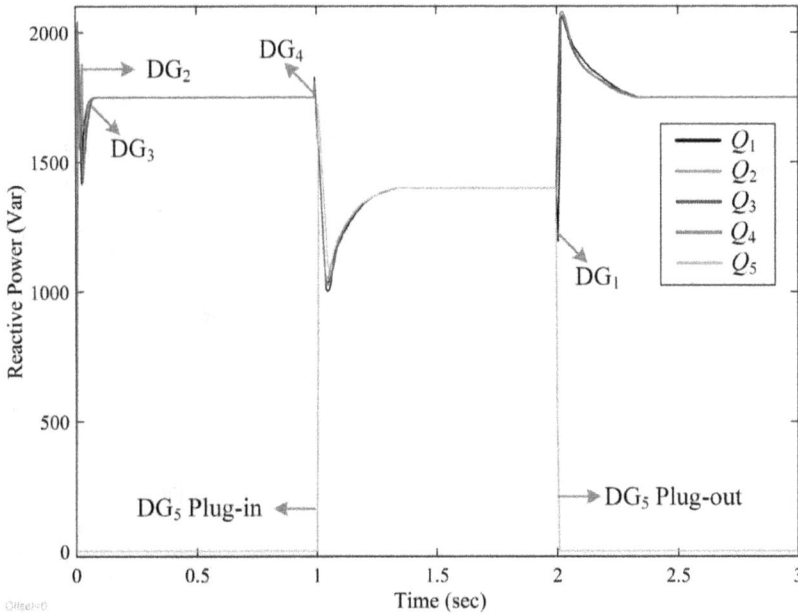

FIGURE 4.6d Reactive power waveforms during PnP functionality by using BP-based API controller.

4.6.1.3 MIT Rule-Based Adaptive Control Scheme

To update the constant gain parameters of the SC during PnP operation, the simulations are performed using an MIT rule-based adaptive controller, as shown in Figure 4.7. When DG_5 is plugged-in or plugged-out from the MG system, a deviation of frequency and voltage waveforms from their references, but the controller restores these variations very rapidly and approximately within 0.1 sec, they come to its steady-state as presented in Figure 4.7a and b. Moreover, the maximum deviation that occurred in case of frequency is 0.04 Hz, while in case of voltage, a maximum noticeable deviation is 6 V. Moreover, during PnP operation, a new power consensus is created, and the controller ensures proportionate power sharing, as presented in Figure 4.7a and b. From the simulation results presented in Figure 4.7, it can be concluded that as compared to the results presented in Figure 4.6 (simulation results of BP-based API), it shows fast dynamic response and high robustness during PnP operation.

4.6.1.4 Lyapunov Function-Based Adaptive Control Scheme

This controller is applied to further improve the performance and dynamic response of the SC during PnP operation. From the output frequency and

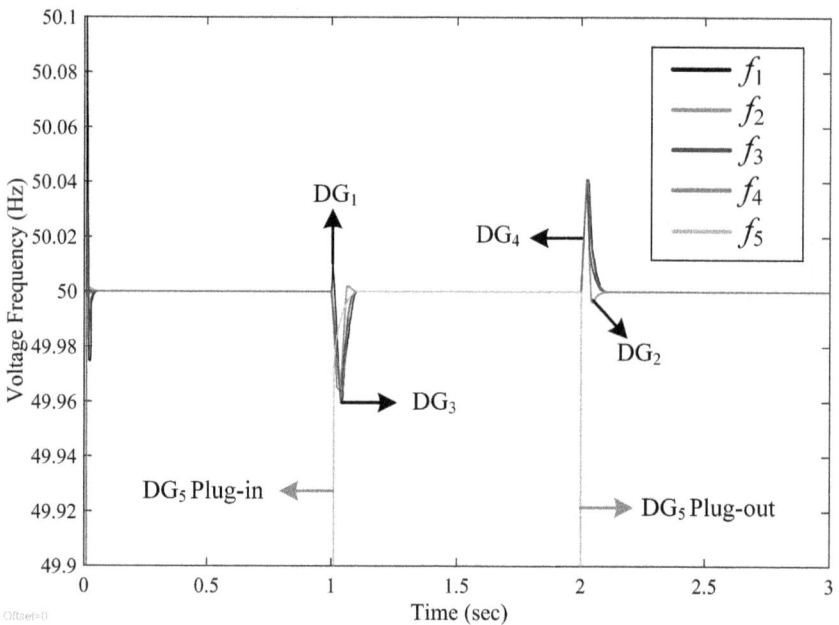

FIGURE 4.7a Frequency waveforms during PnP functionality by using MIT rule-based adaptive controller.

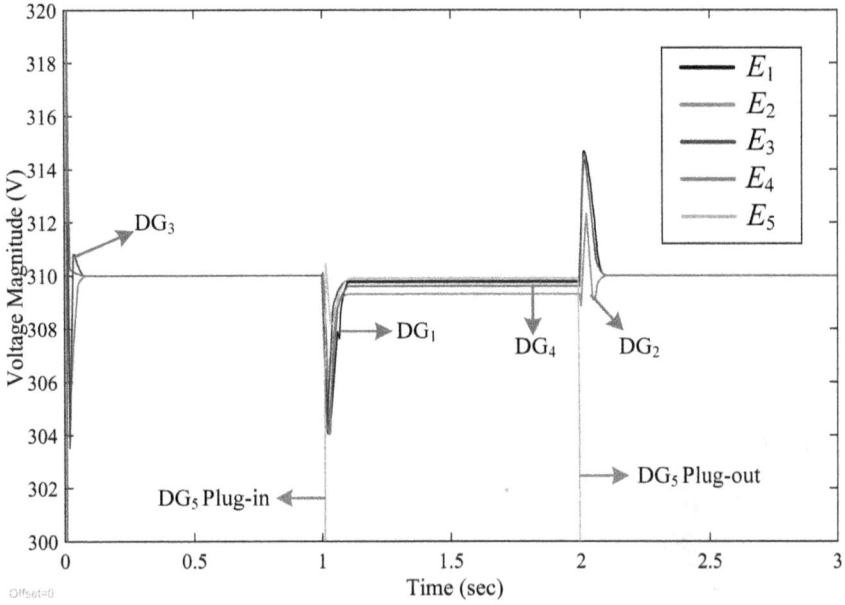

FIGURE 4.7b Voltage waveforms during PnP functionality by using MIT rule-based adaptive controller.

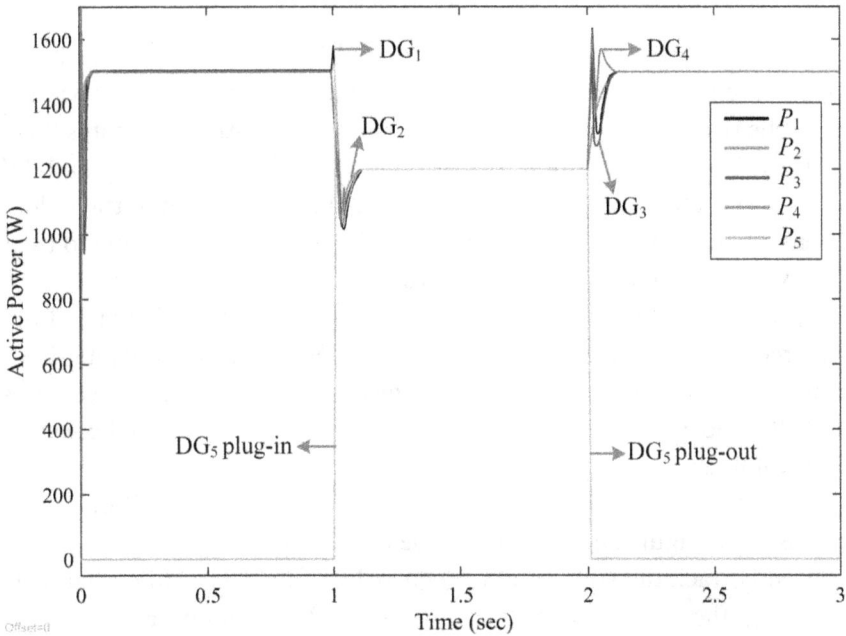

FIGURE 4.7c Active power waveforms during PnP functionality by using MIT rule-based adaptive controller.

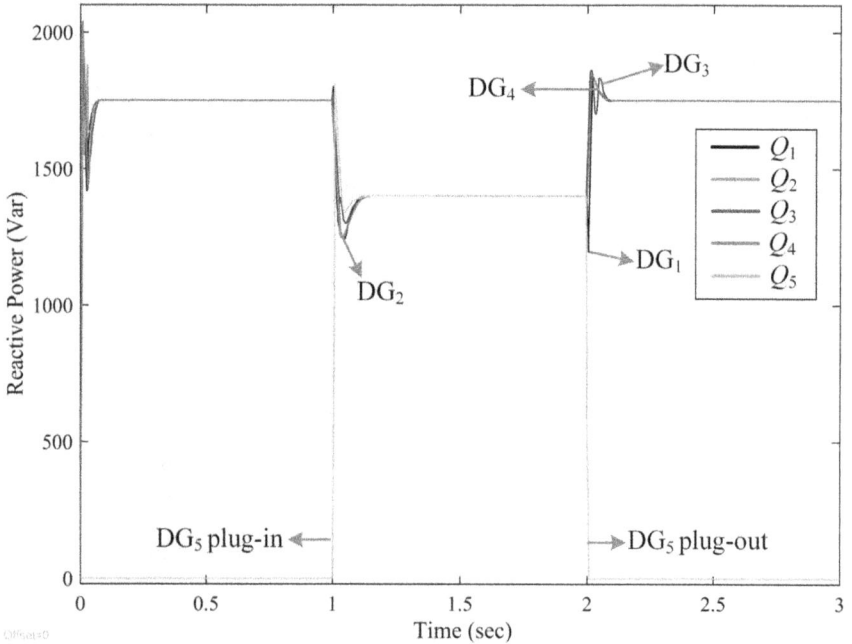

FIGURE 4.7d   Reactive power waveforms during PnP functionality by using MIT rule-based adaptive controller.

voltage waveforms of the DGs as presented in Figure 4.8a and b, it can be observed that when DG_5 is plugged-in into the system, some deviation from the references is observed, but the proposed controller restores them very efficiently. A maximum deviation of 0.015 Hz in frequency and 2 V in voltage is observed, but the controller enables the system to track their references rapidly and accurately, and within 0.07 sec, they come to their steady state. Similarly, at $t=2$ sec, when DG_5 is plugged-out of the system, some deviation in frequency and voltage waveforms is noticed but restored by the controller with high efficacy. Moreover, during plug-in or plug-out operation, a new power consensus is generated in both scenarios and all the load is proportionally shared among the DGs, as can be seen in Figure 4.8c and d.

From the simulation results of the controllers during PnP operation that is from Figures 4.6–4.8, it can be concluded that all the controllers efficiently track the voltages and frequencies of the DGs while accurately updating the power consensus during PnP operation. Furthermore, among the proposed controllers, the best performance is shown by the Lyapunov function-based adaptive controller, which is then followed by MIT rule-based adaptive controller and then the BP-based API controller.

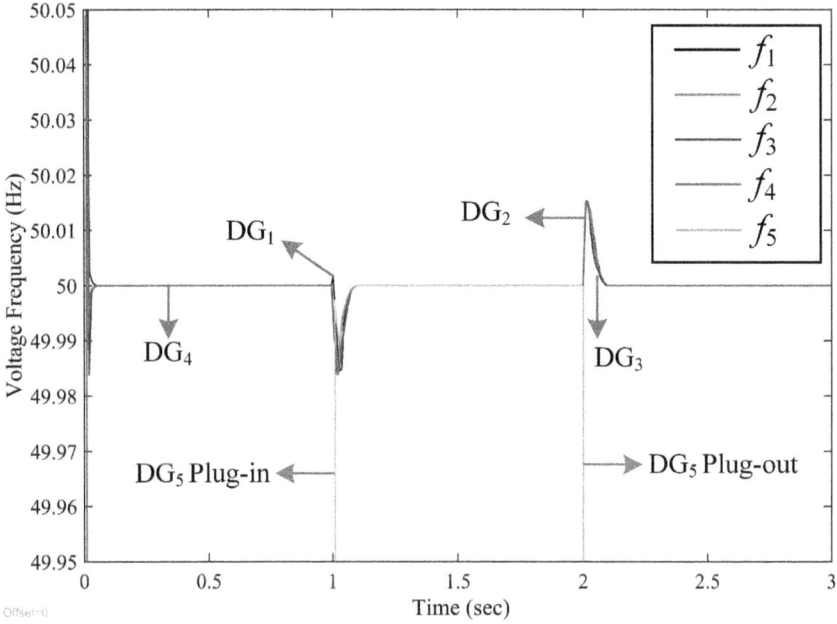

FIGURE 4.8a   Frequency waveforms during PnP functionality by using Lyapunov function-based adaptive controller.

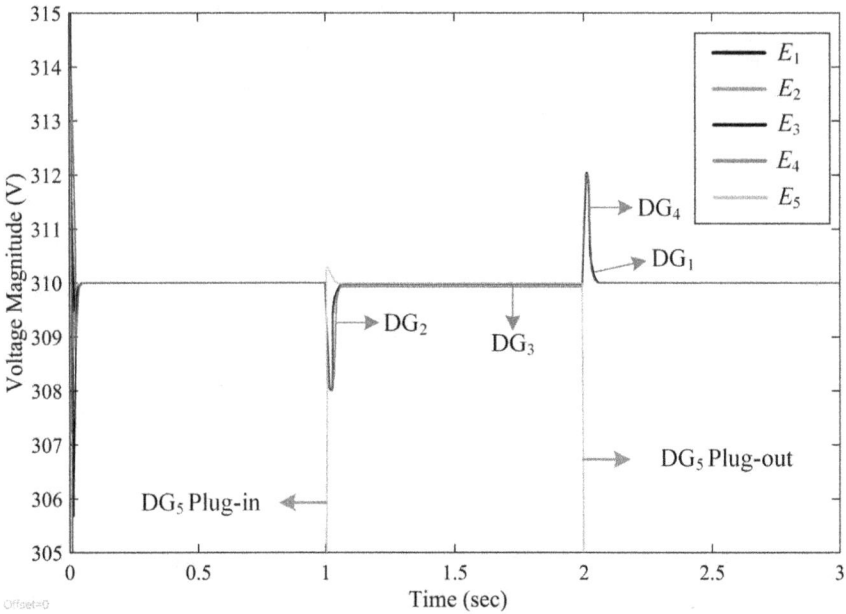

FIGURE 4.8b   Voltage waveforms during PnP functionality by using Lyapunov function-based adaptive controller.

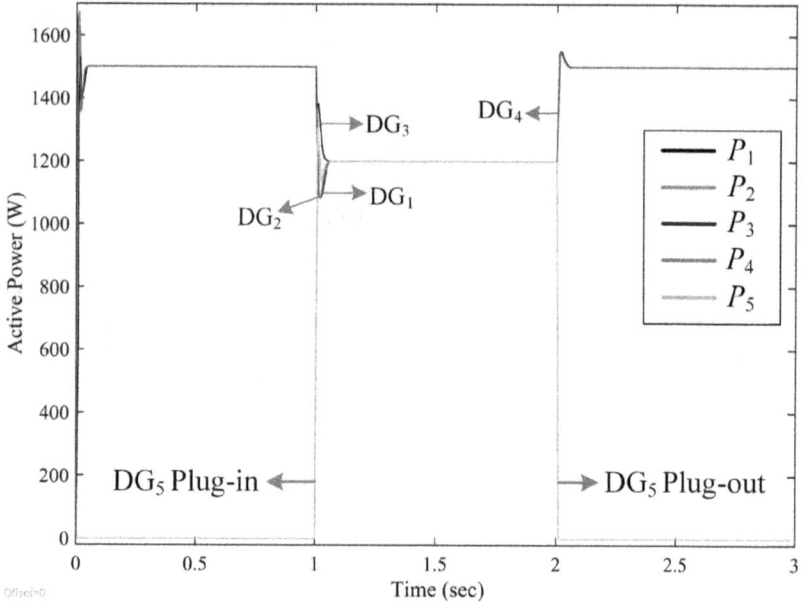

FIGURE 4.8c Active power waveforms during PnP functionality by using Lyapunov function-based adaptive controller.

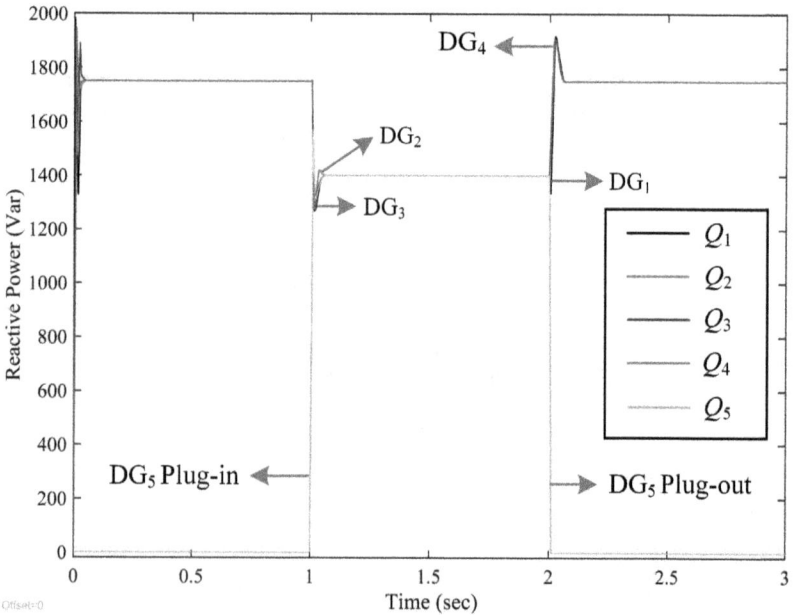

FIGURE 4.8d Reactive power waveforms during PnP functionality by using Lyapunov function-based adaptive controller.

TABLE 4.3   Variation in Load with Respect to Time

Time Period (sec)	Loads (kW+ kVar)		
	L_0	L_1, L_2	L_3, L_4
$0 \leq t \leq 1$	2+3	1+1	1+1
$1 \leq t \leq 2$	2+3	2+2	0.5+0.5
$2 \leq t \leq 3$	2+3	1+1	1+1

4.6.2 Load Variation Condition

Based on consumer activity, a load changes at any instant in an MG; there-fore, the controller must be robust enough to ensure the system's stability in case of any load variation. Hence, to validate the controllers' performance, it is assumed that the load varies at every second according to the pattern shown in Table 4.3. From Table 4.3, it can be seen that during interval $0 \leq t \leq 1$, the values of the loads $L_1 = L_2 = L_3 = L_4 = 1$ kW+1 kVar while $L_0 = 2$ kW+3 kVar. At $t = 1$ sec, the loads changes and during interval $1 \leq t \leq 2$, the values of loads become $L_1 = L_2 = 2$ kW+2 kVar, $L_3 = L_4 = 0.5$ kW+0.5 kVar, and $L_0 = 2$ kW+3 kVar. At $t = 2$ sec, the load varies again and during time interval the values of the load become $L_1 = L_2 = L_3 = L_4 = 1$ kW+1 kVar while $L_0 = 2$ kW+3 kVar.

4.6.2.1 BP-Based API Control Scheme

A BP-based API control scheme is used to improve the performance of the distributed SC presented in Ref. [27] in case of load variation. When the load varies with time according to the pattern shown in Table 4.3, the performance of the controller can be seen in Figure 4.9. From the simula-tion results, it can be seen that when a load variation occurs, the controller restores the frequency and voltage effectively, as presented in Figure 4.9a and b. Although some deviations from their references occur in output waveforms, i.e., in case of frequency regulation, a maximum deviation of 0.32 Hz occurs, while in case of voltage regulation, a maximum devia-tion of 9.8 V is observed, but the BP-based API controller restores these deviations within 0.32 sec. Furthermore, in case of power sharing, at $t = 1$ sec, when the total load changes from 6 kW+7 kVar to 7 kW+8 kVar, a new and updated power consensus is created and the load is distributed homogeneously between all DGs. At $t = 2$ sec, when again a change in load occurs, i.e., load changes from 7 kW+8 kVar to 6 kW+7 kVar, then again a new consensus is created, and the load is shared between all the DGs homogeneously, as presented in Figure 4.9c and d.

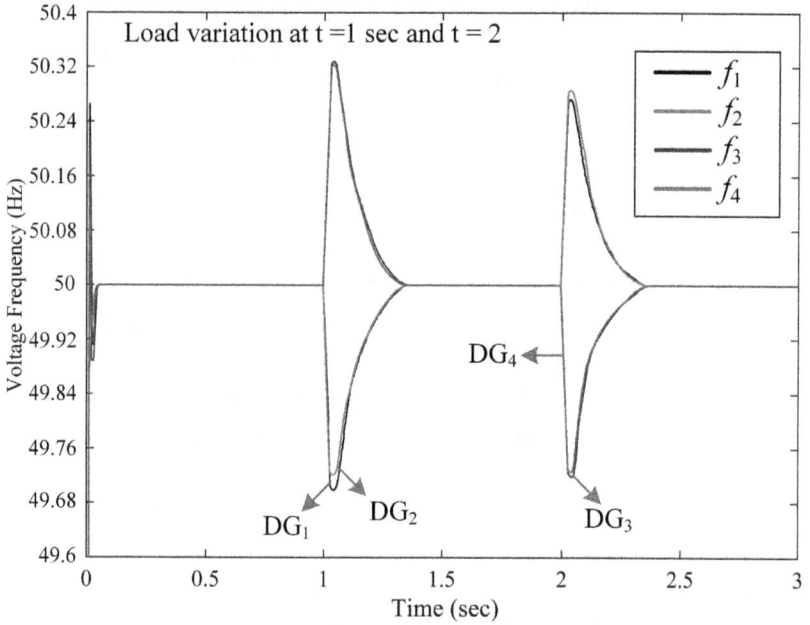

FIGURE 4.9a Frequency waveforms during load variation using BP-based API control scheme.

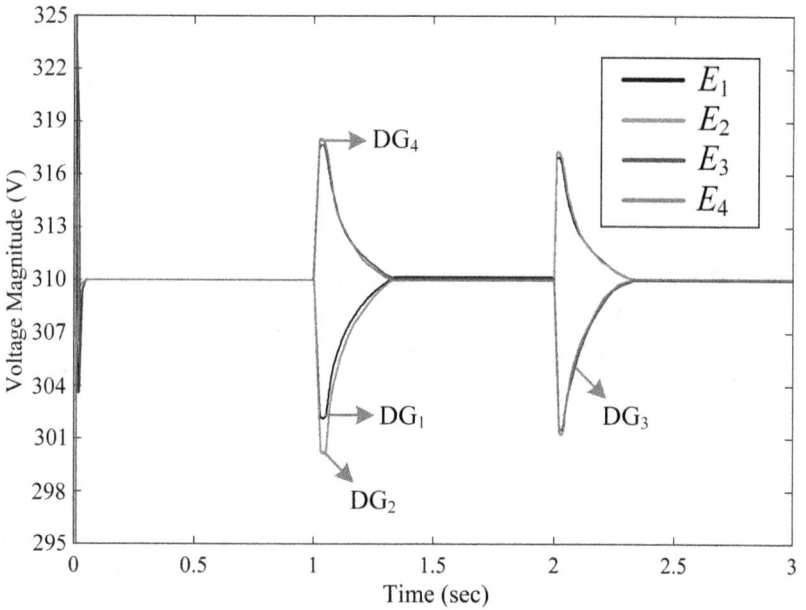

FIGURE 4.9b Voltage waveforms during load variation using BP-based API control scheme.

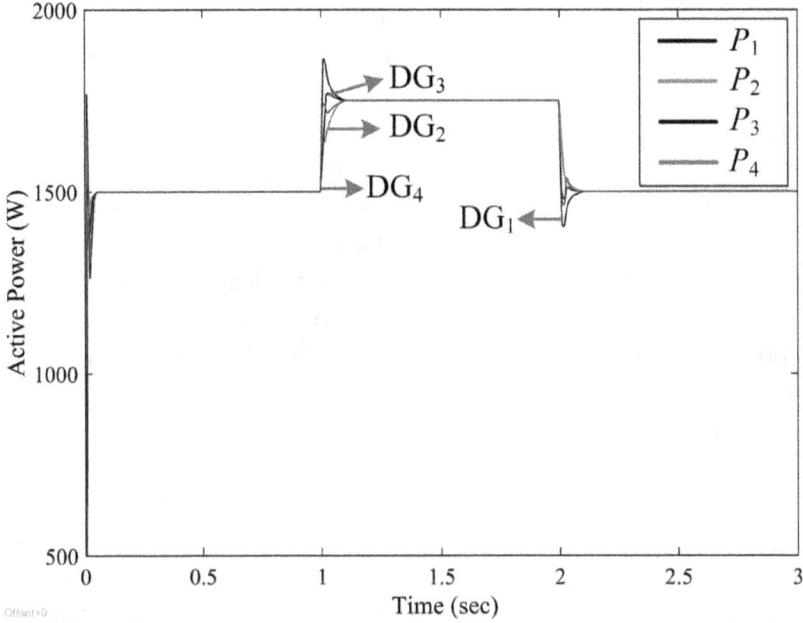

FIGURE 4.9c Active power waveforms during load variation using BP-based API control scheme.

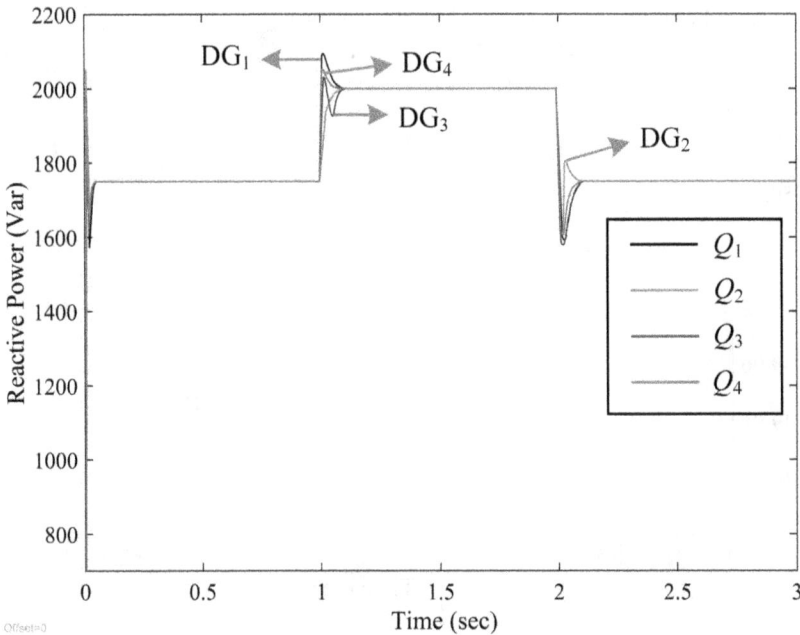

FIGURE 4.9d Reactive power waveforms during load variation using BP-based API control scheme.

4.6.2.2 MIT Rule-Based Adaptive Control Scheme

To validate the performance of an MIT rule-based adaptive distributed SC in case of load variation, simulations are conducted as presented in Figure 4.10. When a variation in load occurs at $t=1$ sec and $t=2$ sec, some deviation from the references in frequency and voltage waveforms, i.e., 0.04 Hz and 4.3 V occurs, respectively. However, the applied control scheme restores the frequency and voltage deviations within 0.12 sec, as shown in Figure 4.10a and b. Moreover, when the load increases at $t=1$ sec, i.e., from 6 kW + 7 kVar to 7 kW + 8 kVar, a power consensus is updated and proportionally sharing the load between the DGs. Likewise, when the total load decreases at $t=2$ sec, a proposed controller ensures a proportionate sharing, as presented in Figure 4.10c and d.

4.6.2.3 Lyapunov Function-Based Adaptive Control Scheme

In case of load variation according to the time intervals presented in Table 4.3, a Lyapunov function-based adaptive control scheme regulate the voltage and frequency of MG very effectively as shown in Figure 4.11a and b.

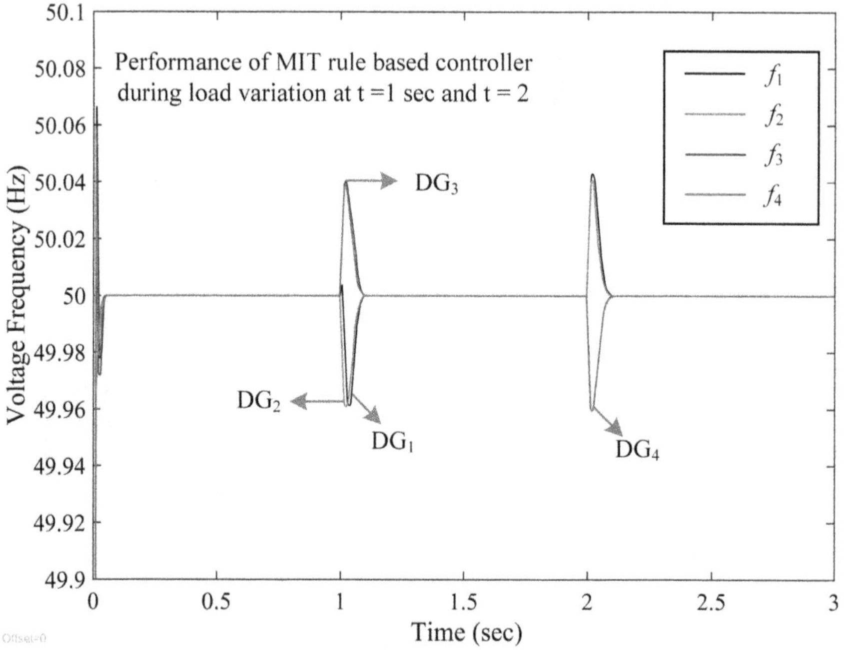

FIGURE 4.10a Frequency waveforms during load variation using an MIT rule-based adaptive control scheme.

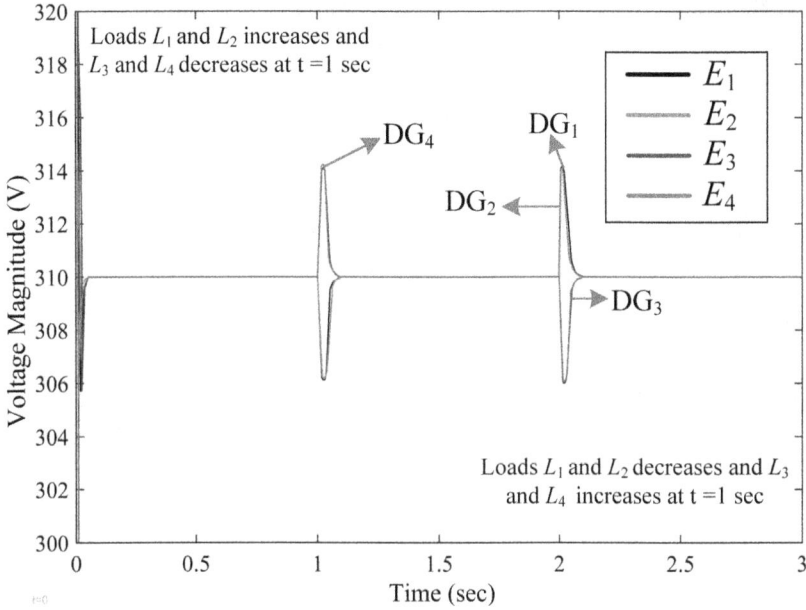

FIGURE 4.10b   Voltage waveforms during load variation using an MIT rule-based adaptive control scheme.

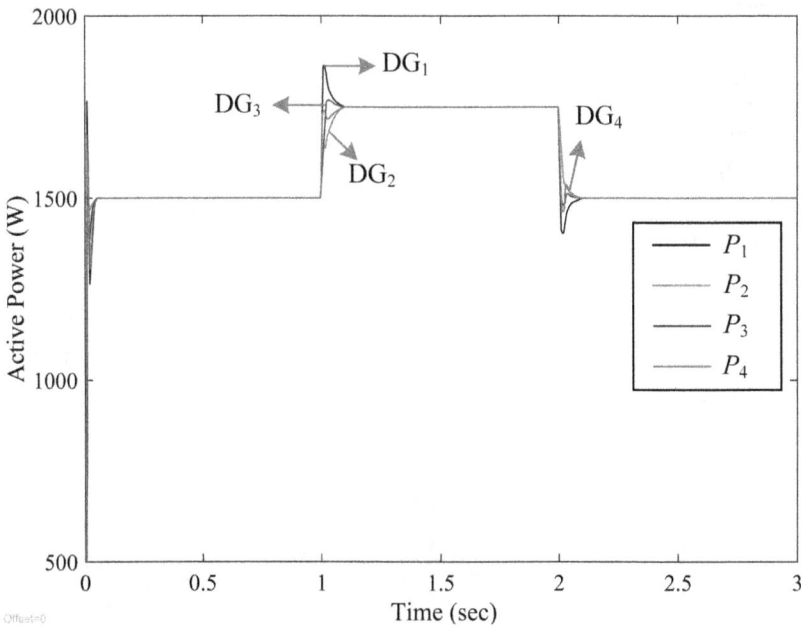

FIGURE 4.10c   Active Power waveforms during load variation using an MIT rule-based adaptive control scheme.

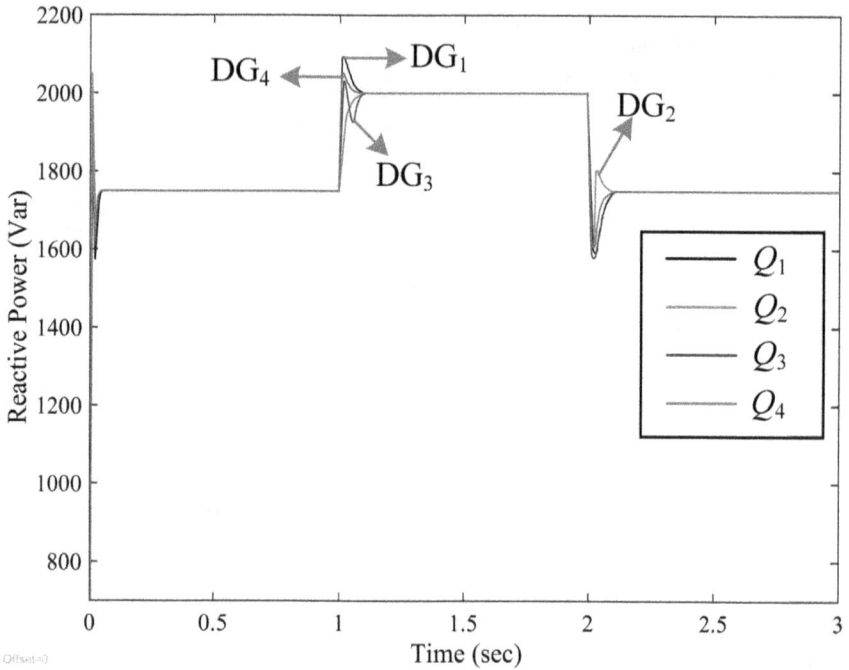

FIGURE 4.10d Reactive power waveforms during load variation using an MIT rule-based adaptive control scheme.

A deviation of 0.02 Hz in frequency and 2 V in voltage waveforms are observed, but the controller restores them efficiently and within 0.07 sec, the waveforms come to its steady-state position. Moreover, when the load changes at $t=1$ sec and $t=2$ sec, a power consensus is updated at both of these instants and accurate sharing of load between the DGs is achieved, as presented in Figure 4.11c and d.

From Figures 4.9 to 4.11, it can be observed that all the controllers accurately track the voltage and frequency reference while achieving proportionate power sharing. From the comparative analysis, it can be concluded that the Lyapunov function-based distributed SC shows a more efficient and fast dynamic response compared with the other two controllers in case of load variation.

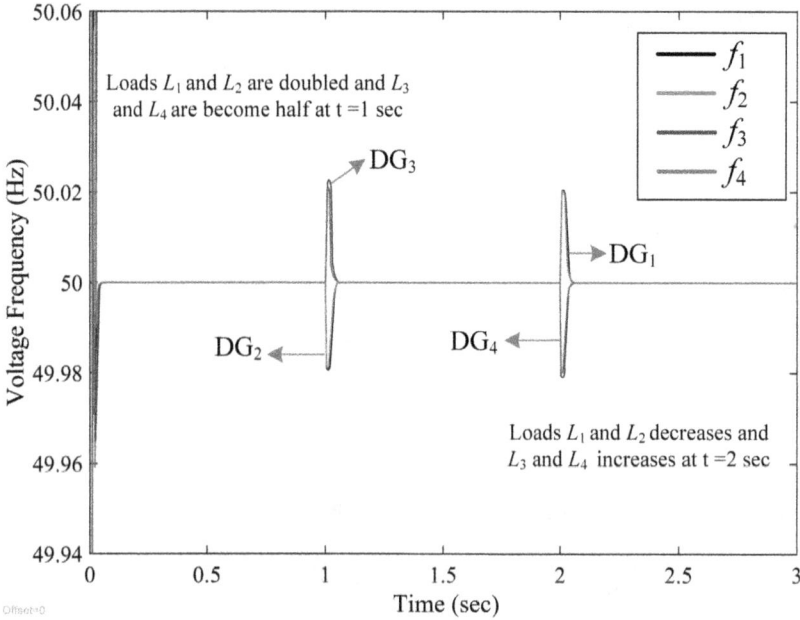

FIGURE 4.11a Frequency waveforms during load variation using Lyapunov function-based distributed SC.

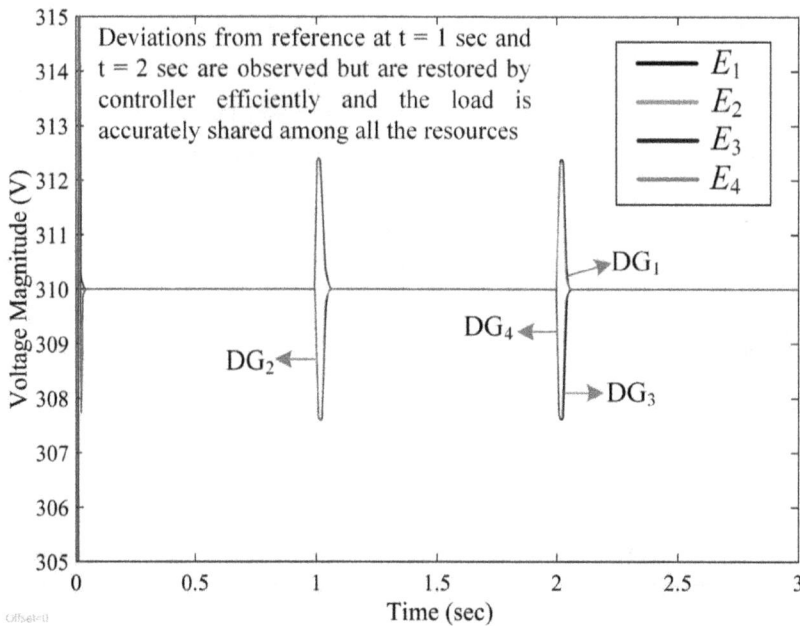

FIGURE 4.11b Voltage waveforms during load variation using Lyapunov function-based distributed SC.

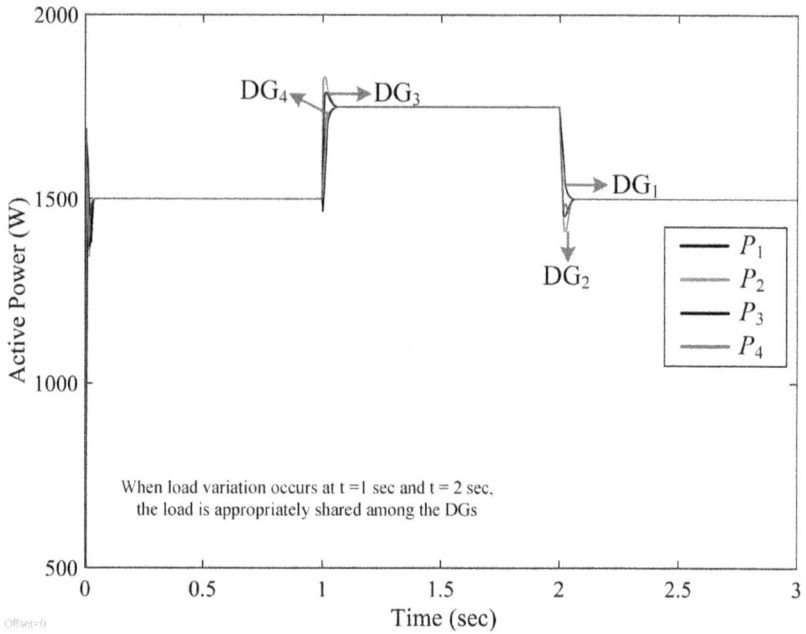

FIGURE 4.11c   Active power waveforms during load variation using Lyapunov function-based distributed SC.

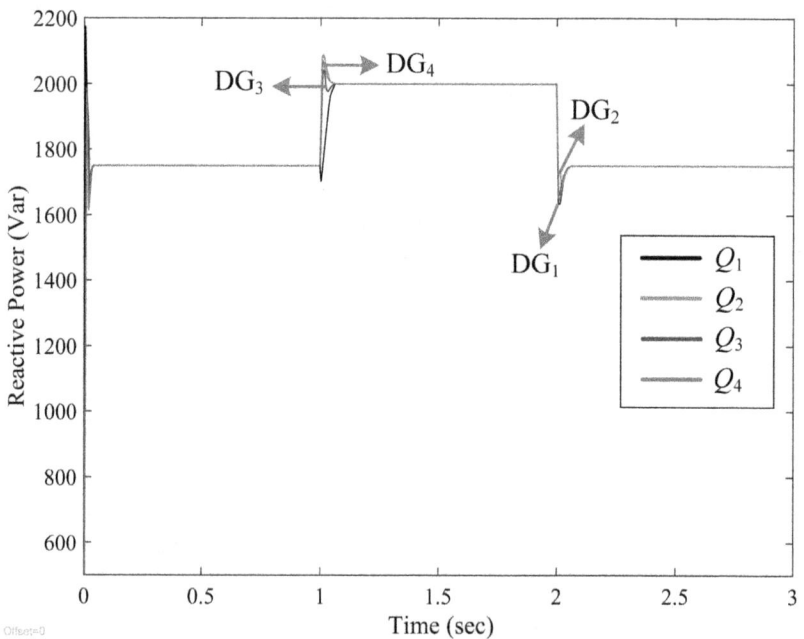

FIGURE 4.11d   Reactive power waveforms during load variation using Lyapunov function-based distributed SC.

4.7 CONCLUSION

The distributed secondary controllers based on adaptive techniques for voltage and frequency regulation and power sharing in IS MG are proposed in this chapter. The proposed SCs that are based on adaptive techniques do not require any prior information about the system dynamics; thus, we can refer to them as model-free controllers. The performance of the proposed controllers is validated through simulations that are performed in different scenarios, such as PnP scenario and load variation conditions. From the simulation results, it can be concluded that high-performance degradation is observed when a distributed SC without any adaptive technique is used. However, when adaptive techniques are used to update the gain parameters of the SC, the performance of the system is significantly enhanced. Moreover, among the proposed adaptive controllers, a Lyapunov-based adaptive controller shows higher performance (low steady-state deviation and fast transient response) than BP-based API and MIT rule-based adaptive controller.

4.8 EXERCISES

1. Calculate the values of proportional gain, resonant gain, and the components of resonant filter by considering DC-input voltage $(V_{DC})=600$ V, resonant frequency $(f_{res})=60$ Hz, and sampling time $(T_s)=0.8\times10^{-5}$ sec while considering the rest of parameters according to Table 4.1. Moreover, perform a simulation to test the performance of the PR controller and resonant filter on these values.

2. Consider 8 DG MG system and check the performance of BP-based API, MIT rule-based adaptive controller, and Lyapunov-based adaptive controller during PnP operation and load variation condition.

3. Compare the performance of conventional distributed controller proposed in Ref. [27] with the controllers presented in this chapter during load variation conditions.

REFERENCES

1. Llanos, Jacqueline, Daniel E. Olivares, John W. Simpson-Porco, Mehrdad Kazerani, and Doris Sáez. "A novel distributed control strategy for optimal dispatch of isolated microgrids considering congestion." *IEEE Transactions on Smart Grid* 10, no. 6 (2019): 6595–6606.
2. Lu, Xi, Shiwei Xia, Guangzeng Sun, Junjie Hu, Weiwei Zou, Quan Zhou, Mohammad Shahidehpour, and Ka Wing Chan. "Hierarchical distributed control approach for multiple on-site DERs coordinated operation in microgrid." *International Journal of Electrical Power & Energy Systems* 129 (2021): 106864.

3. Bidram, Ali, Ali Davoudi, Frank L. Lewis, and Josep M. Guerrero. "Distributed cooperative secondary control of microgrids using feedback linearization." *IEEE Transactions on Power Systems* 28, no. 3 (2013): 3462–3470.

4. Shafiee, Qobad, Josep M. Guerrero, and Juan C. Vasquez. "Distributed secondary control for islanded microgrids—A novel approach." *IEEE Transactions on Power Electronics* 29, no. 2 (2013): 1018–1031.

5. Bidram, Ali, Ali Davoudi, Frank L. Lewis, and Zhihua Qu. "Secondary control of microgrids based on distributed cooperative control of multi-agent systems." *IET Generation, Transmission & Distribution* 7, no. 8 (2013): 822–831.

6. Simpson-Porco, John W., Florian Dörfler, Francesco Bullo, Qobad Shafiee, and Josep M. Guerrero. "Stability, power sharing, & distributed secondary control in droop-controlled microgrids." In *2013 IEEE International Conference on Smart Grid Communications (SmartGridComm)*, pp. 672–677. IEEE, Vancouver, BC, Canada, 2013.

7. Simpson-Porco, John W., Qobad Shafiee, Florian Dörfler, Juan C. Vasquez, Josep M. Guerrero, and Francesco Bullo. "Secondary frequency and voltage control of islanded microgrids via distributed averaging." *IEEE Transactions on Industrial Electronics* 62, no. 11 (2015): 7025–7038.

8. Bidram, Ali, Ali Davoudi, and Frank L. Lewis. "Two-layer distributed cooperative control of multi-inverter microgrids." In *2014 IEEE Applied Power Electronics Conference and Exposition-APEC 2014*, pp. 2364–2371. IEEE, Fort Worth, TX, USA, 2014.

9. Zhao, Changhong, Enrique Mallada, Steven H. Low, and Janusz Bialek. "Distributed plug-and-play optimal generator and load control for power system frequency regulation." *International Journal of Electrical Power & Energy Systems* 101 (2018): 1–12.

10. Tegling, Emma, Martin Andreasson, John W. Simpson-Porco, and Henrik Sandberg. "Improving performance of droop-controlled microgrids through distributed PI-control." In *2016 American Control Conference (ACC)*, pp. 2321–2327. IEEE, Boston, MA, USA, 2016.

11. Lai, Jingang, Xiaoqing Lu, and Xinghuo Yu. "Stochastic distributed frequency and load sharing control for microgrids with communication delays." *IEEE Systems Journal* 13, no. 4 (2019): 4269–4280.

12. Lu, Tianguang, Zhaoyu Wang, Qian Ai, and Wei-Jen Lee. "Interactive model for energy management of clustered microgrids." *IEEE Transactions on Industry Applications* 53, no. 3 (2017): 1739–1750.

13. Xu, Yinliang, Hongbin Sun, Wei Gu, Yan Xu, and Zhengshuo Li. "Optimal distributed control for secondary frequency and voltage regulation in an islanded microgrid." *IEEE Transactions on Industrial Informatics* 15, no. 1 (2018): 225–235.

14. Lou, Guannan, Wei Gu, Yinliang Xu, Ming Cheng, and Wei Liu. "Distributed MPC-based secondary voltage control scheme for autonomous droop-controlled microgrids." *IEEE Transactions on Sustainable Energy* 8, no. 2 (2016): 792–804.

15. Lai, Jingang, Xiaoqing Lu, Xin Li, and Ruo-Li Tang. "Distributed multiagent-oriented average control for voltage restoration and reactive power sharing of autonomous microgrids." *IEEE Access* 6 (2018): 25551–25561.
16. Bouzid, Allal M., Pierre Sicard, Simon Abourida, and Jean-Nicolas Paquin. "Secondary voltage and frequency restoration control of droop-controlled inverter-based microgrids." In *2017 9th IEEE-GCC Conference and Exhibition (GCCCE)*, pp. 1–6. IEEE, Manama, Bahrain, 2017.
17. Lu, Xiaoqing, Xinghuo Yu, Jingang Lai, Yaonan Wang, and Josep M. Guerrero. "A novel distributed secondary coordination control approach for islanded microgrids." *IEEE Transactions on Smart Grid* 9, no. 4 (2016): 2726–2740.
18. Mohiuddin, Sheik M., and Junjian Qi. "Optimal distributed control of AC microgrids with coordinated voltage regulation and reactive power sharing." *IEEE Transactions on Smart Grid* 13, no. 3 (2022): 1789–1800.
19. Godsil, Chris, and Gordon F. Royle. *Algebraic Graph Theory*. Vol. 207. Cham, Switzerland: Springer Science & Business Media, 2001.
20. Schiffer, Johannes, Romeo Ortega, Alessandro Astolfi, Jörg Raisch, and Tevfik Sezi. "Conditions for stability of droop-controlled inverter-based microgrids." *Automatica* 50, no. 10 (2014): 2457–2469.
21. Kundur, Prabha. "Power system stability." *Power System Stability and Control* 10 (2007): 7-1.
22. Guo, Fanghong, Changyun Wen, Jianfeng Mao, and Yong-Duan Song. "Distributed secondary voltage and frequency restoration control of droop-controlled inverter-based microgrids." *IEEE Transactions on Industrial Electronics* 62, no. 7 (2014): 4355–4364.
23. Zhang, Ningyun, Houjun Tang, and Chen Yao. "A systematic method for designing a PR controller and active damping of the LCL filter for single-phase grid-connected PV inverters." *Energies* 7, no. 6 (2014): 3934–3954.
24. Avila, Victor Habermann, and Vicente Leite. "Control of grid-connected inverter output current: A practical review." In *2020 9th International Conference on Renewable Energy Research and Application (ICRERA)*, pp. 232–235. IEEE, Glasgow, UK, 2020.
25. Khan, Muhammad Yasir Ali, Haoming Liu, Salman Habib, Danish Khan, and Xiaoling Yuan. "Design and performance evaluation of a step-up DC–DC converter with dual loop controllers for two stages grid connected PV inverter." *Sustainability* 14, no. 2 (2022): 811.
26. Rosso, Roberto, Xiongfei Wang, Marco Liserre, Xiaonan Lu, and Soenke Engelken. "Grid-forming converters: Control approaches, grid-synchronization, and future trends—A review." *IEEE Open Journal of Industry Applications* 2 (2021): 93–109.
27. Simpson-Porco, John W., Qobad Shafiee, Florian Dörfler, Juan C. Vasquez, Josep M. Guerrero, and Francesco Bullo. "Secondary frequency and voltage control of islanded microgrids via distributed averaging." *IEEE Transactions on Industrial Electronics* 62, no. 11 (2015): 7025–7038.

28. Haroon, Zeeshan, Bilal Khan, Umar Farid, Sahibzada Muhammad Ali, and Chaudhry Arshad Mehmood. "Switching control paradigms for adaptive cruise control system with stop-and-go scenario." *Arabian Journal for Science and Engineering* 44 (2019): 2103–2113.
29. Pankaj, Swarnkar, Jain Shailendra Kumar, and Rajesh Kumar Nema. "Comparative analysis of MIT rule and Lyapunov rule in model reference adaptive control scheme." *Innovative Systems Design and Engineering* 2, no. 4 (2011): 154–162.
30. Khalil, Hassan K. "Nonlinear systems, chapter 4." *ESP* 128 (2002): 111–194.
31. Khan, Muhammad Yasir Ali, Haoming Liu, Jie Shang, and Jian Wang. "Distributed hierarchal control strategy for multi-bus AC microgrid to achieve seamless synchronization." *Electric Power Systems Research* 214 (2023): 108910.
32. Khan, Muhammad Yasir Ali, Haoming Liu, Ren Zhang, Qi Guo, Haiqing Cai, and Libin Huang. "A unified distributed hierarchal control of a microgrid operating in islanded and grid connected modes." *IET Renewable Power Generation* 17, no. 10 (2023): 2489–2511.

CHAPTER 5

Distributed Secondary Controller for Microgrid to Achieve Synchronization

5.1 INTRODUCTION AND LITERATURE REVIEW

In an MG, the energy-generating units produce energy either in the form of AC or DC and are connected to the MG through PE interface (in case of an AC MG, inverters are used). Moreover, an MG has the ability to operate in IS mode or GC mode. Usually, when an MG is operating in a GC mode, an inverter acts as a current-controlled source; on the contrary, when an MG is operating in an IS mode, an inverter acts as a voltage-controlled source. In any of these operating modes, the control of the inverter plays a very important role, as it not only regulates the frequency and voltage and ensures power-sharing but is also responsible for the smooth synchronization of an MG operating in IS mode with the UG.

It is very mandatory for the smooth and uninterruptible operation of the power system that the voltage phase angle (δ), magnitude (E), and frequency (f) on both sides of PCC must be synchronized with each other for initiating a reconnection command. In case of a GC mode, an inverter acts as a current controlled source because the DGs in an MG get the VP support from the UG [1]. However, in case of an IS mode, the UG did

DOI: 10.1201/9781003594284-5

not support an MG; therefore, an inverter that acts as a voltage source converter is responsible for eliminating the VP mismatch from both sides of the PCC and ensuring an accurate power flow [2]. In this case, it is assumed that the VP of the UG is uncontrollable and constant; hence, a voltage source converter is responsible for regulating the VP and ensuring smooth reconnection of IS MG with the UG. If the VP is not synchronized properly, then after a reconnection, high inrush current and transients will appear, which may affect the stability of the system and damage the components that are frequency sensitive [3].

To ensure a smooth and seamless reconnection of IS MG with the UG, numerous researchers have presented different control approaches and techniques, such as the authors in Ref. [82] proposed a droop control technique. This method enables an MG to perform efficiently either in IS mode or in GC mode; however, it is not able to smoothly reconnect an IS MG with the UG. Moreover, as the droop controller is implemented without any CN among the DGs, therefore, this technique is not able to ensure an accurate sharing of load among the DGs [4]. To achieve seamless synchronization, the authors in Ref. [5] proposed a centralized control architecture. In this method, a high-bandwidth CN is required to effectively perform the control objectives, which causes a significant increase in cost. Moreover, by using this controller, an accurate power flow cannot be achieved in GC mode [6].

In Ref. [7], a synchronizer is used to attain synchronization in case of two parallel connected DGs. Although in this method, a VP mismatch from both sides of the PCC is significantly minimized but in case of MG with multi-buses, this method is not very practical. As in today's modern MG system, numerous DGs are integrated with different buses; therefore, it is not a practical solution to synchronize every DG individually with the UG. A control strategy based on PLL is presented in Ref. [8] to achieve seamless synchronization. In this technique, a VP mismatch is removed by using two second-order integrators in $\alpha\beta$ reference frame. In this method, a numerical analysis to derive the synchronization process is not discussed; moreover, the performance of the controller in case of multi-bus system is also not presented. To achieve a seamless reconnection, the authors in Ref. [9] used a self-synchronized method in which virtual synchronization is used to connect an MG with the UG. A CN in this method is based on centralized architecture; therefore, a high-bandwidth CN is required for data acquisition and control that result in high cost. Furthermore, this method is only feasible in the no-load condition.

The authors in Ref. [10] use a droop-based method to attain a seam-less reconnection. In this approach, only the frequency and phase angle synchronization is considered while neglecting the synchronization of voltage magnitude. The frequency and phase angle oscillations are signifi-cantly reduced, but in GC mode, a power flow between UG and MG is not realized properly. In Ref. [11], PI controllers along with PLLs are used to achieve synchronization. In this approach, the frequency and δ-mismatch is detected using PLLs and the generated error is then given to PI control-lers to generate the correction signals for the droop controller to perform the synchronization process. Although, the synchronization is achieved, but this method lacks to explain how smoothly the process is performed; moreover, it also does not ensure appropriate power flow. A synchroniza-tion is attained by using a droop controller, but there are some challenges that cannot be met such as (a) it pushes the frequency and voltage magni-tude away from their references, (b) degradation in performance in case of any uncertainty, and (c) cannot achieve appropriate power flow. Due to these limitations of droop controller, it automatically calls for the SC level to cope with these challenges.

A synchronization technique in which a SC level is implemented in a centralized architecture is proposed in Refs. [12,13]. As discussed in Chapter 3, a centralized control architecture requires a high-bandwidth CN to perform the control actions as every DG is required to be con-nected to one central controller; moreover, it also shows high sensitivity against single point failure. Due to these limitations of centralized con-troller, a distributed controller is more preferably used in MG synchroni-zation applications. The authors in Ref. [14] uses a predictive controller in a distributed architecture that predict the magnitude and phase angle of voltage to attain a synchronization. In IS mode, phase angle of the DGs output waveforms are used to control the frequency. In a GC mode, the voltage magnitude and phase angle of the DGs are controlled through PI controller while avoiding the regulation of frequency. This unbounded and un-regulated frequency may cause system instability and may affect the performance of the components that are frequency sensitive. A virtual oscillator control-based control strategy to attain a synchronization is pro-posed in Ref. [15]. This method attains seamless synchronization and also improves the operation of the MG, but the initial non-linear characteris-tics and fast response of virtual oscillator may affect the system's stability. The authors in Ref. [16] proposed a distributed architecture-based control system in which three separate regulators are used to control the voltage, active power, and reactive power. This system ensures appropriate sharing of power and improves system performance, but it is not able to completely eliminate the VP mismatch at a node that is near to PCC.

A leader-follower concept base distributed controller to attain a seamless reconnection is presented in Ref. [17–19]. In this technique, a primary droop controller is used along with a novel distributed SC law that enhances a system operation and performance. Moreover, a supervisory controller is also designed at a tertiary level for data acquisition and generation and transmission of correction compensation signals to a leader-DG. A leader-DG is then responsible for transmitting the correction signals to the follower-DGs to adjust their VP at PCC. This technique shows very effective performance; however, a main drawback of this method is that it is unable to resolve a natural conflict between phase and frequency regulation. Similarly, a pinning consensus-based distributed SC is proposed in Refs. [20,21]. All the three components of the VP, i.e., phase, magnitude, and frequency are controlled explicitly. In this method, all the DGs of the MG system are responsible for regulating the frequency while only a selected DG termed as Regulating-DG (R-DG) is used to restore the phase angle and magnitude mismatch. Although, a smooth reconnection is ensured in this method but it lacks to discuss the generation of compensation signals for the SC of the R-DG. Moreover, it also lack to discuss the control mechanism of the static switch. Furthermore, different synchronization methods discussed are summarized in Table 5.1.

TABLE 5.1　Comparison of Different Synchronization Methods

Ref.	Controller	CS	VP			RCBFP	APF	PnP	CSG	BS
			f	δ	E					
[7]	Traditional synchronizer	M	×	✓	✓	×	×	×	×	SBS
[10]	Droop	H	✓	✓	×	×	×	×	×	MBS
[11]	Droop	M	✓	✓	✓	×	×	×	✓	MBS
[13]	Centralized	L	✓	×	✓	×	✓	✓	×	MBS
[14]	Distributed	H	×	✓	✓	×	✓	×	✓	MBS
[15]	Distributed	H	×	✓	✓	×	✓	×	×	MBS
[16]	Distributed	H	✓	✓	✓	✓	✓	×	✓	MBS
[19]	Distributed	H	✓	✓	✓	×	✓	×	✓	MBS
[20]	Distributed	H	✓	✓	✓	✓	✓	×	×	MBS
P	Distributed	H	✓	✓	✓	✓	✓	✓	✓	MBS

Abbreviations: APF, Appropriate Power Flow; BS, Bus System; CS, Convergence Speed; CSG, Correction Signal Generator; H, High; M, Medium; P, Proposed; RCBFP, Resolve the Conflict between Frequency and phase; S, Slow.

5.2 CONTRIBUTIONS

From the above-discussed state-of-the-art literature review, it can be concluded that the distributed system must have (a) the ability to eliminate the VP mismatch from the PCC of an MG and the UG to attain seamless synchronization, (b) successfully resolve a conflict between frequency and phase regulation, (c) the controller should be converged synchronously and rapidly, and (d) must be able to handle different kinds of uncertainties (PnP, load variation, etc.). On the basis of these points, a pinning consensus-based distributed controller is designed in this research work to attain a seamless reconnection of IS MG with the UG. The contributions of this work can be explained as follows [22,23]:

- All the DGs in an MG are connected with their neighbouring DGs through a sparse CN, and all are responsible for regulating their frequencies and output powers according to their capacity. Moreover, only an R-DG is responsible for eliminating the magnitude and phase angle mismatch from both sides of the PCC, i.e., MG side and the UG side.

- In this method, the phase angle, frequency, and voltage magnitude are controlled explicitly, thus enabling us to resolve the phase angle and frequency regulation conflict without compromising the system's reliability.

- To avoid current surges and ensure smooth synchronization, PI controller, FPI controller, and SD-based FLC are designed. These controllers continuously measure the dynamic behaviour of the system and use adjustable parameters to generate the correction signals for the SC in order to ensure a seamless reconnection. Moreover, the small signal stability of the controller is also discussed that verifies the design criteria of the control parameters.

5.3 SYSTEM CONFIGURATION

In this section, physical-cyber network used in this work is explained. Moreover, this section also discusses the grid synchronization method.

5.3.1 Physical-Cyber Model

An MG that is considered in this work contains n number of buses that can either be DGs or loads. Every DG consists of an inverter, DC source, and filter. Furthermore, it is considered that c_{th}-DG is integrated with the

PCC through lines that are mainly inductive and have a reactance of X_c. Thus the injected active and reactive power of c_{th}-DG at PCC can be presented as [24]:

$$P_c = \frac{E_{PCC}E_c}{X_c}\sin(\theta_c) \qquad (5.1)$$

$$Q_c = \frac{E_{PCC}E_c\cos(\theta_c) - E_{PCC}^2}{X_c} \qquad (5.2)$$

where P_c and Q_c presents the injected active and reactive of c_{th}-DG at PCC; E_{PCC} and E_c shows the voltage magnitude at PCC and c_{th}-DG, respectively; $\theta_c = \delta_c - \delta_{PCC}$ in which the phase angle of output voltage of c_{th}-DG is presented by δ_c and the phase angle of voltage at PCC is denoted by δ_{PCC}; and X_c presents the reactance such that $X_c = 1/\sum_{c=1}^{n}(1/X_{cd})$.

A communication between the DGs in an MG is done through meshed CN where every DG behaves like an agent and can be modelled using a digraph as $G = (v, \xi, A)$. A set of vertices (agents or DGs tagging) is presented by v, such that $v = \{1, 2, \dots, n\}$; a set of edges is denoted by ξ, and $\xi \in [v] \times [v]$. If nodes c and d share information with each other then $(a_c, a_d) \in \xi$ thus G is considered to be undirected. $A = \mathbb{R}^{|N| \times |N|}$ presents an adjacent matrix whose elements (a_{cd}) presents the connection weight; it means that if the edge exists between nodes c and d then the elements $a_{cd} = a_{dc} > 0$, hence when there is no edge between node c and d then $a_{cd} = a_{dc} = 0$. If the nodes are organized in a way that the edges of successive pair are connected in an order form, then a path will exist between node c and d, i.e., $\mho(cd) \in [v] \times [v]$, and hence, it can be said that the graph is connected. Thus, under consensus the dynamics of the DG can be presented as [25]:

$$\dot{i}_c = -\sum_{c,d \in v} a_{cd}(i_c - i_d) \qquad (5.3)$$

where i_c denotes the state of c_{th}-DG. An expression presented in (5.3) needed the information of the states which are locally accessible to DGs, i.e., the state i_c is accessed by c_{th}-DG. On the contrary, when the states are

not accessed locally by the DGs, then a pinning consensus-based algorithm is used and can be expressed as:

$$i_c = -\sum_{c,d\in v} a_{cd}(i_c - i_d) - j_c \qquad (5.4)$$

where j_c denotes the control input of the pinning algorithm and can be
expressed as:

$$j_c = \alpha_c [g(h_c) - \tilde{g}] \qquad (5.5)$$

where α_c presents the gain of the weighted pinning; $g(h_c)$ represent the
states that is not locally accessed by the DGs and is regulated by i_c; and $\tilde{g}$
represents its desired value. Furthermore, if c_{th}-DG is pinned then only
$\alpha_c > 0$, otherwise $\alpha_c = 0$. When the system reaches its desired consensus,
then the state converges as:

$$\lim_{t\to\infty}(i_c - i_d) = 0 \qquad (5.6)$$

5.3.2 Description of Grid Synchronization Method

A general schematic of multi-bus MG system (consisting of transmission
lines, Grid Forming DGs (GF-DGs), R-DGs, loads, and physical-cyber network) integrated with the UG through SS as presented in Figure 5.1. From
Figure 5.1, it can be seen that the MG (i.e., PCC of an MG) is integrated with
the UG through a static switch; hence, it is very important that the VP at
the PCC must be synchronized with the UG before initiating a GC process.
In order to attain a smooth synchronization process, the VP of both sides
of SS, i.e., PCC side and UG side is measured and the difference or error is
to the Synchronous Controller (SYNC). The SYNC is responsible for eliminating the VP difference and generating the correction signal (phase angle
$(\Delta\delta^s_{ref})$ and voltage magnitude (ΔE^s_{ref})). These generated signals are then fed
to a distributed SC in order to perform the required control actions. If an
MG is re-synchronized with the UG without proper synchronization, it will
cause a high inrush current that may lead to system instability and damage
the frequency-sensitive components. Therefore, proper synchronization is
very important to ensure system stability and limit the inrush current.

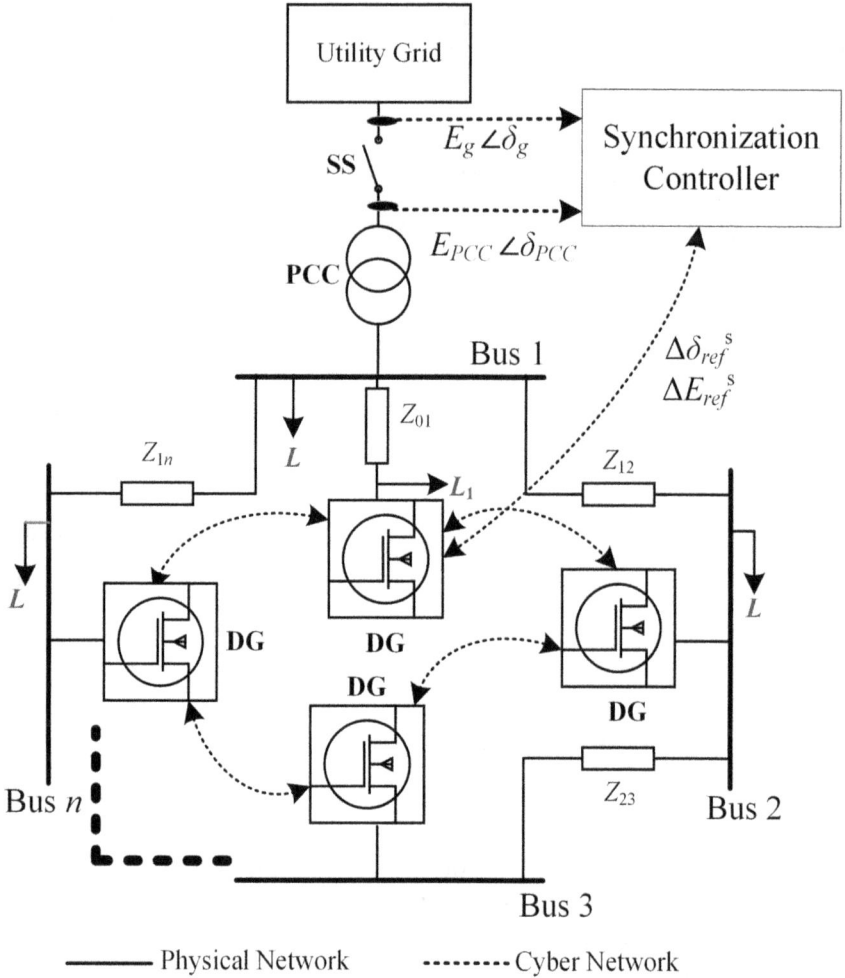

FIGURE 5.1   Generalized schematic of an MG.

5.4 FREQUENCY REGULATION AND ACTIVE POWER SHARING CONTROLLER

To eliminate the phase angle and frequency mismatch and ensure proportionate power sharing a distributed SC can be presented as:

$$\omega_c = \omega_{\text{ref}} - m_c P_c + \varpi_c \qquad (5.7)$$

$$\frac{d\varpi_c}{dt} = \mathbb{Z}_f \left[\sum_{d=1}^{n} a_{cd}(\varpi_d - \varpi_c) - (\omega_c - \omega_{\text{ref}}) - \gamma_c \Delta \delta_{\text{ref}}^s \right] \qquad (5.8)$$

where ω_c is the angular frequency of c_{th}-DG $(c=1,2,\ldots,n)$; ω_{ref} is the nominal angular frequency, respectively; m_c are the $(P_c-\omega_c)$ droop coefficient; ϖ_c is the SC variable; $\mathbb{Z}_f$ is the gain of the SC; if there exists a cyber-link among c_{th}-DG and d_{th}-DG then $a_{cd}=1$, however, on the contrary, when there is no link then $a_{cd}=0$; γ is the gain for δ regulation such that if c_{th}-DG is selected as R-DG then $\gamma_c=1$, hence for the GF-DGs $\gamma=0$; $\Delta\delta^s_{ref}$ is the correction signal generated by the SYNC. Moreover, an adjacent matrix can be presented as $A=\{a_{cd}\}$. A detailed schematic of the distributed SC is shown in Figure 5.2.

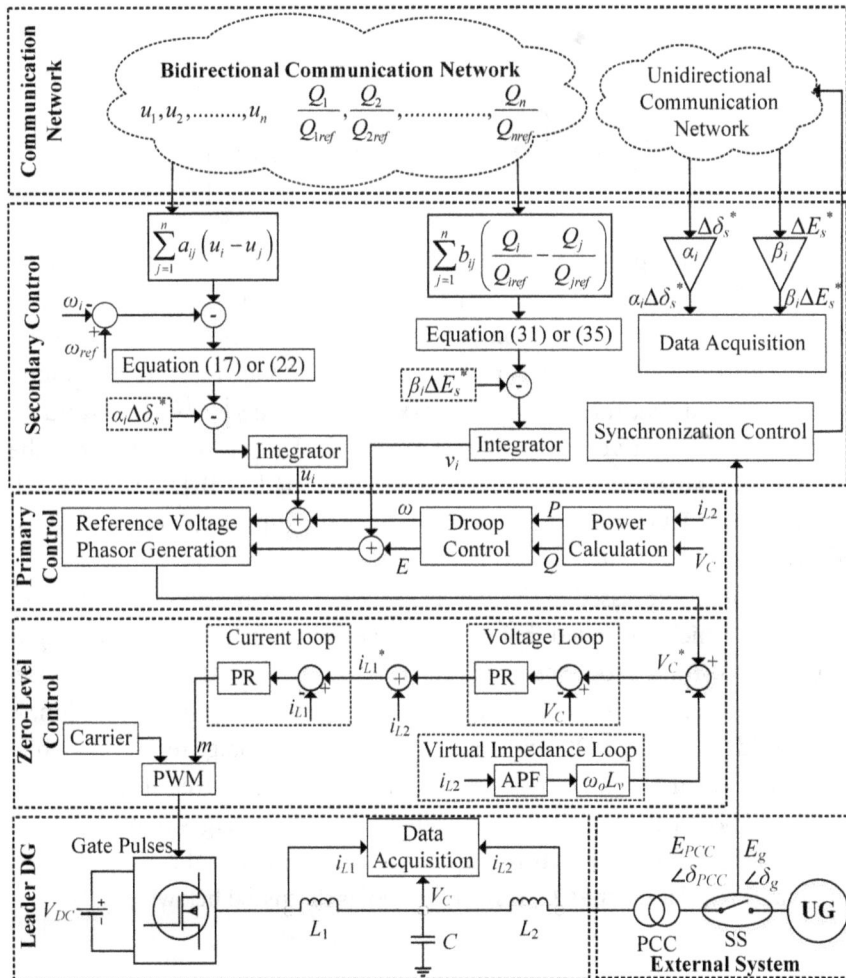

FIGURE 5.2 Schematic of distributed controller.

In a steady state, the left side of (5.8) becomes zero, and the following declarations can be made: (a) in accordance with the nominal frequency, all DGs adjust their frequencies, i.e., $\omega_c = \omega_d = \omega_{ref}$ and frequency mismatch is eliminated [26], (b) δ mismatch from both sides of SS is removed, and (c) same droop shift and an appropriate active power sharing are attained by all DGs, i.e., $\overline{\omega}_c = \overline{\omega}_d$.

5.5 VOLTAGE REGULATION AND REACTIVE POWER-SHARING CONTROLLER

A voltage regulation and reactive power sharing distributed SC can be presented as:

$$E_c = E_{ref} - n_c Q_c + V_c \tag{5.9}$$

$$\frac{dv_c}{dt} = \mathbb{Z}_E \left[\sum_{d=1}^{n} b_{cd} \left(\frac{Q_d}{Q_d^*} - \frac{Q_c}{Q_c^*} \right) - \alpha_c \Delta E_{ref}^s \right] \tag{5.10}$$

where E_c is the output voltage of c_{th}-DG; E_{ref} is the nominal voltage; n_c is the $(Q_c - E_c)$ droop coefficients; v_c is the SC variable; $\mathbb{Z}_E$ is the gain of the SC; if there exists a cyber-link among c_{th}-DG and d_{th}-DG then $b_{cd} = 1$, however, on the contrary, when there is no link then $b_{cd} = 0$, however, for simplicity it is considered that the adjacent matrixes $A = B = \{a_{cd}\} = \{b_{cd}\}$; Q_c^* and Q_d^* represents the reactive power of c_{th}-DG and d_{th}-DG, respectively; α is the gain for E-regulation such that if c_{th}-DG is selected as R-DG then $\alpha_c = 1$, hence, for GF-DGs $\alpha = 0$; ΔE_{ref}^s is the voltage magnitude correction signal generated by the SYNC.

An equation expressed in (5.10) is a case of the distributed SC presented in Ref. [26] in which it is stated that c_{th}-DG can either contribute to voltage regulation $(\alpha_c \neq 0, b_{cd} = 0)$ or in reactive power sharing $(\alpha_c = 0, b_{cd} \neq 0)$; otherwise, a steady-state error appears in both regulators. Based on this formulation, it can be said that a single DG will have to contribute to voltage regulation while other DGs ensure accurate reactive power sharing. Moreover, if the impedances and power flow in the line are small, then the voltages of other nodes will gather around it [27]. In this research work, the same methodology is used for voltage regulation at PCC. A related proof is discussed below:

Proof

Let consider that $[1]_{n,m}$ presents $n \times m$ matrix having all entries are one. If the expression in (5.10) converges then it can be written as:

$$\sum_{d=1}^{n} b_{cd}\left(\frac{Q_c}{Q_c^*} - \frac{Q_d}{Q_d^*}\right) = -\alpha_c \Delta E_{\text{ref}}^s \qquad (5.11)$$

As discussed above, there must be at least one DG that is involved in voltage regulation; thus, an R-DG is selected in this prospect, thus $[\alpha] = \text{diag}(\alpha_1, \alpha_2, \ldots, \alpha_n)\,[1]_{n,1}$, where $\alpha_c \geq 0$. Hence, in matrix form (5.11) can be presented as:

$$L[Q] = -[\alpha]\Delta E_{\text{ref}}^s \qquad (5.12)$$

where L denotes a Laplacian matrix. Hence, (5.12) can be given as:

$$L\left[\hat{Q}\right] + L\left[\Delta \hat{Q}\right] = -[\alpha]\Delta E_{\text{ref}}^s \qquad (5.13)$$

where $L\left[\hat{Q}\right] = \text{avg}\left(Q_c/Q_c^*\right)\cdot[1]_{n,1}$ and the steady-state error in reactive power sharing can be given as $\left[\Delta\hat{Q}\right] = [Q] - \left[\hat{Q}\right] = \left(\Delta\hat{Q}_1, \Delta\hat{Q}_2, \ldots, \Delta\hat{Q}_n\right)$. Let's assume that $L\left[\Delta\hat{Q}\right] = 0$ then (5.13) can be given as:

$$L\left[\hat{Q}\right] = -[\alpha]\Delta E_{\text{ref}}^s \qquad (5.14)$$

Multiply (5.14) by $[1]_{n,1}$, we get:

$$L\left[\hat{Q}\right][1]_{n,1} = -[\alpha]\Delta E_{\text{ref}}^s [1]_{n,1} \qquad (5.15)$$

From (5.15), it can be easily proved that $L[1]_{n,1} = 0$ hence, $-[\alpha]\Delta E_{\text{ref}}^s [1]_{n,1} = 0$. As discussed above $[\alpha]$ is a non-zero vector, thus $[Q] = \left[\hat{Q}\right]$ and $\Delta E_{\text{ref}}^s = 0$. It can be concluded that the convergence of the controller will occur in a steady-state without introducing any error either in reactive power sharing or voltage regulation.

5.6 STABILITY ANALYSIS

A small signal model is discussed to present the design of the controller and the stability of the system. The inverter output controller at the primary control level has a very fast transient response; therefore, it can be ignored, according to [28]. Furthermore, the system's frequency has fast convergence; therefore, the impacts of delays in the neighbouring DG are ignored. However, the delays in voltage regulation are modelled as first-order low-pass filters. Moreover, it is assumed that in case of any variation in UG frequency occurs, it should be restricted to its nominal value.

Based on the above assumptions, an expression in (5.7) for c_{th}-DG can be modelled as:

$$\dot{\delta}_1 = \omega_c - \omega_{ref} = -m_c P_c + \omega_c \qquad (5.16)$$

Put (5.1) in (5.16) would give us:

$$\dot{\delta}_1 = \omega_c - m_c \left(\frac{E_{PCC} E_c}{X_c} \sin(\delta_c - \delta_{PCC}) \right) \qquad (5.17)$$

Considering (5.10), c_{th}-DG can be modelled as:

$$\omega_{LP}^{-1} \dot{E}_c = E_{ref} - E_c - n_c Q_c + v_c \qquad (5.18)$$

where ω_{LP} denotes the cut-off angular frequency if low pass filter. By putting (5.2) in (5.18) we get:

$$\omega_{LP}^{-1} \dot{E}_c = E_{ref} - E_c - n_c \left(\frac{(E_c - E_{PCC})E_{PCC}}{X_c} \cos(\delta_c - \delta_{PCC}) \right) + v_c \qquad (5.19)$$

During the modelling process it is assumed E_{PCC} and E_c are constant and $\sin(\delta_c - \delta_{PCC}) \approx (\delta_c - \delta_{PCC})$. Furthermore, when Q and P are mainly dependent on voltage and phase angle, respectively, then only a droop controller is valid [29]. Moreover, a maximum allowable variation in voltage from its reference is ±5% [30], hence, $E_c \approx E_d$, $E_{ref} \approx E_{PCC}$, and $\cos(\delta_c - \delta_{PCC}) \approx 1$. On the basis of these considerations (5.17) and (5.19) can be presented in (5.20) and (5.21), respectively, as:

$$\dot{\delta}_1 = \omega_c - m_c \left(\frac{E_{PCC} E_c}{X_c} (\delta_c - \delta_{PCC}) \right) \qquad (5.20)$$

$$\omega_{LP}^{-1} \dot{E}_c = E_{PCC} - E_c - n_c \left(\frac{(E_c - E_{PCC})E_{PCC}}{X_c} \right) + v_c \qquad (5.21)$$

As Q and P are decoupled from each other therefore to facilitate the design their stability is presented separately.

5.6.1 Phase Angle/Frequency Stability Analysis

Let consider that $\mathbf{K}_c = E_{PCC} E_c / X_c$ and put in (5.20), we get:

$$\dot{\delta}_c = \varpi_c - m_c \mathbf{K}_c \left(\delta_c - \delta_{PCC} \right) \qquad (5.22)$$

As a VP at PCC of an MG, it depends on the VP of the DGs, thus $\delta_{PCC} = \sum \kappa_c \delta_c$; where κ_c is a constant and can be given as $\kappa_c = X_P / X_c$ whereas $X_P = \left(\sum (1/X_c) \right)^{-1}$. Moreover, some matrices that are used in this analysis can be defined as: 0_n denotes a zero matrix, i.e., all entries are zero; 1_n denotes an $n \times n$ identity matrix; $\gamma = \text{diag}(\gamma_c)$; $\kappa = \text{diag}(\kappa_c)$; $\mathbf{K} = \text{diag}(K_c)$; and $\mathbf{m} = \text{diag}(m_c)$. Moreover, if L is the Laplacian matrix, then $1_n + L$ is a positive definite [25]. On the basis of these assumptions (5.22) can be expressed in matrix form as:

$$\dot{\delta} = \varpi + \mathbf{m} \mathbf{K} (1_n - \kappa) \delta \qquad (5.23)$$

where δ represents the system's operating state. Considering (5.8), it can be presented in a matrix form as:

$$\dot{\varpi} = \mathbb{Z}_f \mathbf{m} \mathbf{K} (1_n - \kappa) \delta - \mathbb{Z}_f (1_n + L) \varpi - \gamma \kappa \delta \qquad (5.24)$$

By combining (5.23) and (5.24), we get a small signal model as:

$$\dot{g}_1 = H_1 g_1 \qquad (5.25)$$

where g_1 denotes a state of the system and H_1 presents the matrix of the system and can be given as:

$$g_1 = \begin{bmatrix} \delta & \varpi \end{bmatrix}^T \qquad (5.26a)$$

$$H_1 = \begin{bmatrix} H_1 & 0 \end{bmatrix} \qquad (5.26b)$$

$$H_1 = \begin{bmatrix} -\mathbf{m} \mathbf{K} (1_n - \kappa) & 1_n \\ \mathbb{Z}_f \mathbf{m} \mathbf{K} (1_n - \kappa) - \gamma \kappa & -\mathbb{Z}_f (1_n + L) \end{bmatrix} \qquad (5.26c)$$

According to (5.26), if the Eigen values of H_1 are in left side in the graph then a system is said to be an exponentially stable. Moreover, to drive the exponentially stability conditions for H_1, Schur complement is used as:

$$\det(s1_n - H_1) = \det\left(s1_n + Z_f(1_n + L)\right)$$

$$\det\left[\left(s1_n + mK(1_n - \kappa)\right) - 1_n\left(s1_n + Z_f(1_n + L)\right)^{-1}\left(Z_f mK(1_n - \kappa) - \gamma\kappa\right)\right]$$

$$(5.27)$$

Let's assume that $x_{0-1} = Z_f mK(1_n - \kappa) - \gamma\kappa$ and $x_{1-1} = mK(1_n - \kappa) + Z_f(1_n + L)$. Put these values in (5.27) would yield us to:

$$\det(s1_n - H_1) = \det\left(s1_n + Z_f(1_n + L)\right)\det\left(s^2 1_n + x_{1-1}s + x_{0-1}\right) \quad (5.28)$$

From (5.28), as $1_n + L$ is a positive definite, then the roots of $\det\left(s1_n + Z_f(1_n + L)\right) = 0$ satisfies $\mathbb{R}(s) < 0$. Moreover, according to Routh-Hurwitz stability criteria, the roots of $\det\left(s^2 1_n + x_{1-1}s + x_{0-1}\right)$ satisfies $\mathbb{R}(s) < 0$, if $\lambda_{\min}\left(x_{1-1} + x_{1-1}^T\right) > 0$ and $\lambda_{\min}\left(x_{0-1} + x_{0-1}^T\right) > 0$. Hence, it provides enough conditions for the exponential stability of the system.

5.6.2 Voltage Regulation Stability Analysis

Let consider that $T_c = E_{PCC}/X_c$ and put in (5.21), we get:

$$\omega_{LP}^{-1}\dot{E}_c = E_{PCC} - E_c - n_c T_c(E_c - E_{PCC}) + v_c \quad (5.29)$$

Conferring to Millman's theorem, $E_{PCC} = \sum U_c E_c$ whereas ρ_c is a constant and can be given as $U_c = (E_c/E_{PCC})\kappa_c$ for $c = 1, 2, \ldots, n$. Moreover, some matrices that are used in this analysis can be defined as: $n = \text{diag}(n_c)$; $\alpha = \text{diag}(\alpha_c)$; $T = [1]_N \text{diag}(T_c)$; and $U = [1]_N \text{diag}(\mathbb{Q}_c)$. On the basis of these assumptions (5.29) can be expressed in matrix form as:

$$\omega_{LP}^{-1}\dot{E} = v - E - nT(1_n - U)E \quad (5.30)$$

where E represents the system's operating state. Considering (5.10), it can be presented in a matrix form as:

$$\dot{v} = -Z_E LT(1_n - U)E - \alpha UE \quad (5.31)$$

By combining (5.30) and (5.31), we get a small signal model as:

$$\dot{g}_2 = H_2 g_2 \tag{5.32}$$

where g_2 denotes a state of the system and H_2 presents the matrix of the system and can be given as:

$$g_2 = \begin{bmatrix} E & v \end{bmatrix}^T \tag{5.33a}$$

$$H_2 = \begin{bmatrix} H_2 & 0 \end{bmatrix} \tag{5.33b}$$

$$H_2 = \begin{bmatrix} \omega_{LP}\left[1_n - nT(1_n - U)\right] & \omega_{LP}1_n \\ -\mathbb{Z}_E LT(1_n - U - U\alpha) & 0 \end{bmatrix} \tag{5.33c}$$

To drive the exponentially stability conditions for H_2, Schur complement ca be given as:

$$\det(s1_n - H_2) = \det(s1_n - 0_n)$$

$$\det \begin{bmatrix} \left(s1_n - \omega_{LP}\left\{-1_n - nT(1_n - U)\right\}\right) - \omega_{LP}1_n(s1_n - 0_n)^{-1} \\ \left(-\mathbb{Z}_E LT(1_n - U - U\alpha)\right) \end{bmatrix} \tag{5.34}$$

Let's assume that $x_{1-2} = \omega_{LP}\left\{1_n - nT(1_n - U)\right\}$ and $x_{0-1} = -\mathbb{Z}_E LT\left(1_n + U(1_n + \alpha)\right)$. Put these values in (5.34), and solve it would give us:

$$\det(s1_n - H_2) = \det(s1_n - 0_n)\det(s^2 1_n + x_{1-2}s + x_{0-2}) \tag{5.35}$$

According to the stability criteria of Routh-Hurwitz, the roots of $\det(s^2 1_n + x_{1-2}s + x_{0-2})$ satisfies $\mathbb{R}(s) < 0$, if the following conditions become true:

$$\lambda_{\min}\left(x_{1-2} + x_{1-2}^T\right) > 0 \tag{5.36a}$$

$$\lambda_{\min}\left(x_{0-2} + x_{0-2}^T\right) > 0 \tag{5.36b}$$

A (5.36) gives sufficient conditions for the system to be exponentially stable.

5.7 SYNCHRONIZATION CONTROL STRATEGIES

To ensure a seamless reconnection and eliminate the voltage and phase angle mismatch between the PCC of the MG and the UG, three different controllers for correction signal generation are presented that are discussed below.

5.7.1 Proportional Integral Controller

A PI controller is used to generate $\Delta\delta_{\text{ref}}^s$ and ΔE_{ref}^s which are then fed to the SC so that it can perform its control tasks effectively. A generalized PI controller can be presented as:

$$\Delta A = k_p e(t) + k_i \int e(t)dt \tag{5.37}$$

where $\Delta A = \Delta\delta_{\text{ref}}^s$ or $\Delta A = \Delta E_{\text{ref}}^s$ is the output of controller; k_p and k_i are the proportional and integral gains, respectively. In this technique, the values from both sides of SS are measured and an error is generated. The error is then fed to a PI controller in order to minimize by using k_p and k_i gains. The output of the controllers are then fed to SC through CN so that the SC perform its mandatory actions efficiently.

5.7.2 Fuzzy Proportional Integral Controller

From a PI control scheme presented in (5.37) it can be observed that the values of k_p and k_i are not adaptive; hence, to improve the performance of the PI controller and make the gains adaptive such that they update in every sampling period, a FLC is used. An FLC consists of fuzzification, Fuzzy Rule (FR) base, inference mechanism, and defuzzification processes. During fuzzification, linguistic terms and membership functions are used to convert the collected crisp inputs into fuzzy sets. During the FR base process, the fuzzy sets are compared with the predefined rules. This comparison is observed in the inference mechanism, and based on the condition, an appropriate rule is selected. Finally, in the defuzzification process, data is converted into a crisp form that is used to update the gains of the PI controller. A generalized schematic of the proposed FLC-based PI controller is presented in Figure 5.3.

The FR that used to update the gains of the PI controller are shown in Table 5.2. Moreover, a fuzzification and defuzzification processes that uses a Gaussian Membership Function (GMF) can be given as:

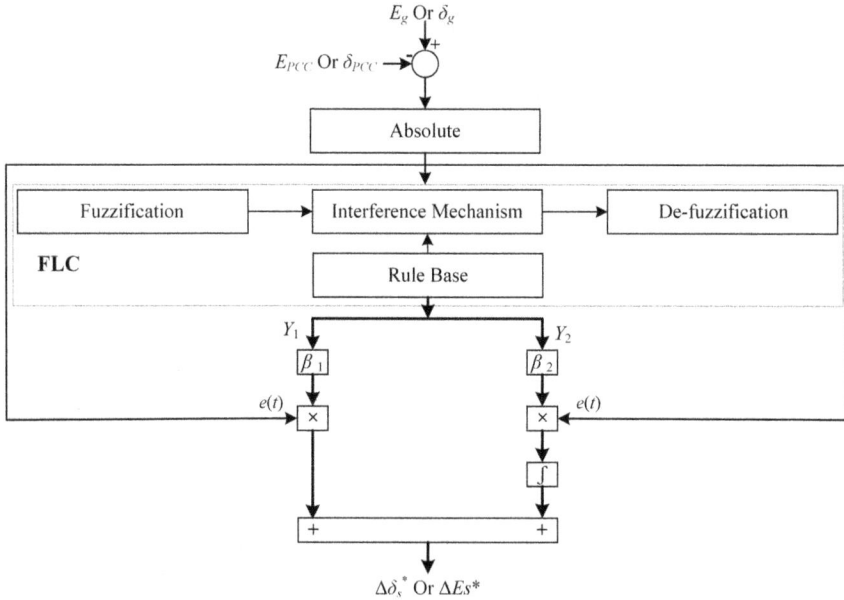

FIGURE 5.3   FLC-based PI controller.

TABLE 5.2   Fuzzy Rules of the Controller

Input Membership Function		Output Membership Function		If-then Rules	
Linguistic Terms	Range	Linguistic Terms	Range	If $|e(t)|$	Then (k_p, k_i)
Large	[0.8, 1.0]	Large	[0.8, 1.0]	Large	Large
Zero	[0, 0.2]	Zero	[0, 0.2]	Zero	Zero
Small	[0.3, 0.7]	Small	[0.3, 0.7]	Small	Small

$$\mu(y) = \exp\left(-\frac{1}{2}\left(\frac{y_c - c_c}{\sigma_c}\right)^2\right) \tag{5.38}$$

where c_c represents the centre and σ_c denotes the variance. Consider a PI controller in (5.37), a proposed FLC-based PI can be mathematically presented as:

$$\Delta A = Y_1 \beta_1 e(t) + Y_2 \beta_2 \int e(t) dt \tag{5.39}$$

where Y_1 and Y_2 are the FLC outputs; β_1 is the learning rate of k_p; and β_2 is the learning rate of k_i.

5.7.3 Steepest Descent-Based FLC

To generate the compensation signal for the elimination of voltage and phase angle mismatch an SD-based FLC is presented. This controller does not need a precise mathematical model and parameters of the system and is given as:

$$i_j = i_{j-1} - \varsigma y_j \tag{5.40}$$

where i_j and i_{j-1} represent the updated and previous value, respectively; ς is the learning rate; y_j denotes a current value.

In this work, an SD algorithm is used to update the output variance, centre, and GMF of FLC to offer an improved solution. A cost function of FLC optimized by SD method is presented as:

$$f(x) = \frac{1}{2} \sum_{c=1}^{m} d_j^2(x) \tag{5.41}$$

where $x = \begin{bmatrix} x_1 & x_2 & x_3 \dots x_n \end{bmatrix}^T \in \mathbb{R}^{n \times 1}$; d_j is a residual function from $\mathbb{R}^n \to \mathbb{R}$ and a residual vector can be given as $d(x) = [d_1(x), d_2(x), \dots, d_m(x)]$. Thus, (41) can be rewritten as:

$$f(x) = 0.5 \|d(x)\|^2 \tag{5.42}$$

- **Design of a Controller**

  To minimize the error this controller uses a first derivative and is able to optimize both non-linear and liner functions [31]. An error function emerged from the cost function can be presented as:

$$e_Z = \frac{1}{2}(x_Z - x^*)^2 \tag{5.43}$$

where e_Z denotes an error; x_Z represent the output of the controller; and x^* shows the reference for the controller. To defuzzify the output during defuzzification, a centre of gravity method is utilized and is presented as:

$$x_Z = \frac{\sum_{c=1}^{R} \Omega_c b_c(x)}{\sum_{c=1}^{R} \Omega_c(x)} \tag{5.44}$$

where Ω_c shows the GMF and can be given as:

$$\Omega_c(x) = \prod_{d=1}^{n} \exp\left(-0.5\left(\frac{y_d^m - c_d^c}{\sigma_d^c}\right)^2\right)$$

(5.45)

where y_d^m presents the crisp value; c_d^c and σ_d^c represent the centre and spread of GMF.

- **Adaptation of Output GMF**

  To minimize the error and make the output of GMF adaptive, take the time derivative of e_z with respect to b_c; we get:

$$\frac{\partial e_z}{\partial b_c} = \vartheta \frac{\partial \vartheta}{\partial b_c} = \vartheta \frac{\partial}{\partial b_c} \left[\frac{\prod_{c=1}^{\mathbb{R}} b_c \exp\left(-0.5\left(\frac{y_d^m - c_d^c}{\sigma_d^c}\right)^2\right)}{\prod_{c=1}^{\mathbb{R}} \exp\left(-0.5\left(\frac{y_d^m - c_d^c}{\sigma_d^c}\right)^2\right)} \right]$$

(5.46)

Hence, according to (5.40) the output GMF can be updated as:

$$b_c(j) = b_c(j-1) - \vartheta_\varsigma \frac{\partial}{\partial b_c} \left[\frac{\prod_{c=1}^{\mathbb{R}} b_c \exp\left(-0.5\left(\frac{y_d^m - c_d^c}{\sigma_d^c}\right)^2\right)}{\prod_{c=1}^{\mathbb{R}} \exp\left(-0.5\left(\frac{y_d^m - c_d^c}{\sigma_d^c}\right)^2\right)} \right]$$

(5.47)

- **Adaptation of Centre**

  Most of the values of GMF transpired at the centre, and the rest of the values are equally spread from the centre. Therefore, for the centre adaptation, take a time derivative of e_z with respect to $c_c(j)$ and after the solution put the values in (5.40) would give us:

$$c_c(j) = c_c(j-1) - \vartheta_\varsigma \left[\vartheta \left(\frac{\sum_{c=1}^{\mathbb{R}} b_c - x_z}{\sum_{c=1}^{\mathbb{R}} \Omega_c(y_d^m, j)} \right) \left(\frac{y_d^m - c_d^c}{\sigma_d^c} \right) \Omega_c(y_d^m, j) \right]$$

(5.48)

- **Adaptation of Variance**

    From (5.45), it can be concluded that the inverse is inversely proportional to GMF magnitude, i.e., if GMF magnitude increases, then variance decreases and vice versa. Therefore, to make a variance adaptive, take the derivate of (5.43) with respect to $\sigma_c(j)$ and put the solution in (5.40) would give us:

$$\sigma_c(j) = \sigma_c(j-1) - \vartheta_\varsigma \left[\vartheta \left(\frac{\left(\sum_{c=1}^{\mathbb{R}} b_c - x_Z \right)}{\sum_{c=1}^{\mathbb{R}} \Omega_c(y_d^m, j)} \right) \left(\frac{y_d^m - c_d^c}{\sigma_d^c} \right) \Omega_c(y_d^m, j) \right]$$

(5.49)

The updated values of variance, centre, and output GMF are then passed by the four components of FLC as discussed above to minimize the error function, thus making a conventional FLC more efficient and robust.

5.8 CONTROL OF STATIC SWITCH AND SYNCHRONIZATION CRITERIA

The main purpose of this research is to eliminate the VP difference between MG and UG to achieve seamless synchronization. Besides generating the correction signals for the SC, a SYNC is also responsible for sending a command to a static switch to perform a closing or opening operation. Once synchronization is achieved, there are some standards that are defined by IEEE-1547 that must be fulfilled before sending a triggering command to a static switch [29]. However, with the advancement of technology and desire to ensure safe operation of the power system, the authors in Ref. [19] presented more strict standards for IS MG reconnection, as presented in Table 5.3. Hence, based on the standards defined in

TABLE 5.3   Standards of Synchronization

Standards	Magnitude Difference (%)	Phase angle difference (°)	Frequency difference (Hz)	DG power rating (kVA)
IEEE Standard 1547–2003 [29]	3	10	0.1	>1,500–10,000
	5	15	0.2	>500–1,500
	10	20	0.3	0–500
[19]	< 0.01 pu	< 5	< 0.05	Criteria used

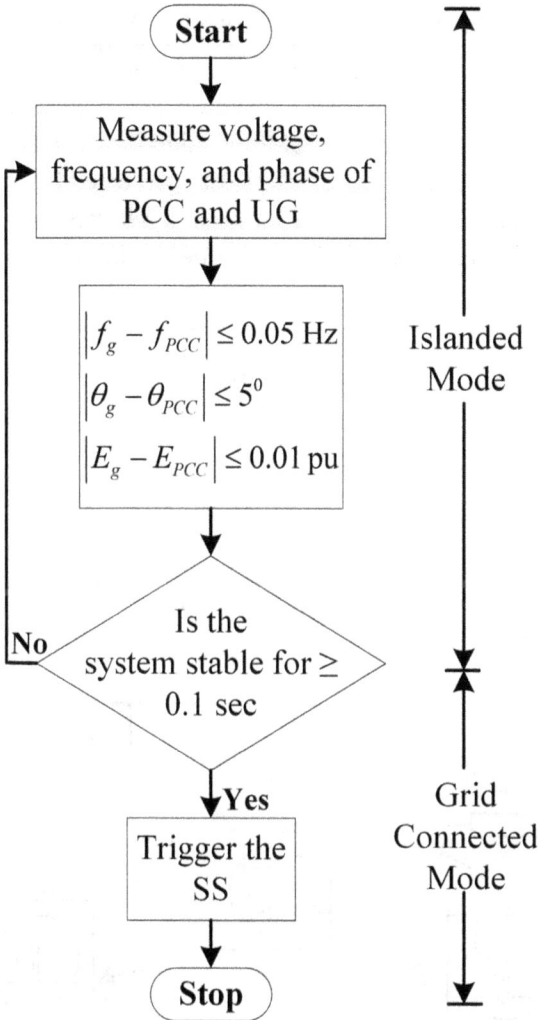

FIGURE 5.4   Schematic of SS triggering.

Ref. [19], the authors in this research work presented a control strategy for static switch triggering as presented in Figure 5.4.

In this method, the VP at PCC of MG and UG is measured and the difference is calculated at every sampling period, and the difference is compared with the predefined limits. If the difference between the VP becomes lower than the set limits and attains it for more than 0.1 sec, then an algorithm will send the triggering command to the static switch, and it will change its state from open to close. On the contrary, if the set limits are not attained for more than 0.1 sec, then the algorithm will take no action until the triggering conditions are fulfilled.

5.9 DISCUSSION AND PERFORMANCE ANALYSIS

To validate the efficacy and performance of the distributed SC, simulations are conducted in MATLAB®/SimPower® environment. An MG system under consideration consists of four radial buses, 4-DGs (DG_1-DG_4), 4 local loads (L_1-L_4), and 1 public load (L_0), as presented in Figure 5.5. In an MG system that is under consideration, DG_1 is selected as an R-DG while all the other DGs are the GF-DGs. The values of different parameters and

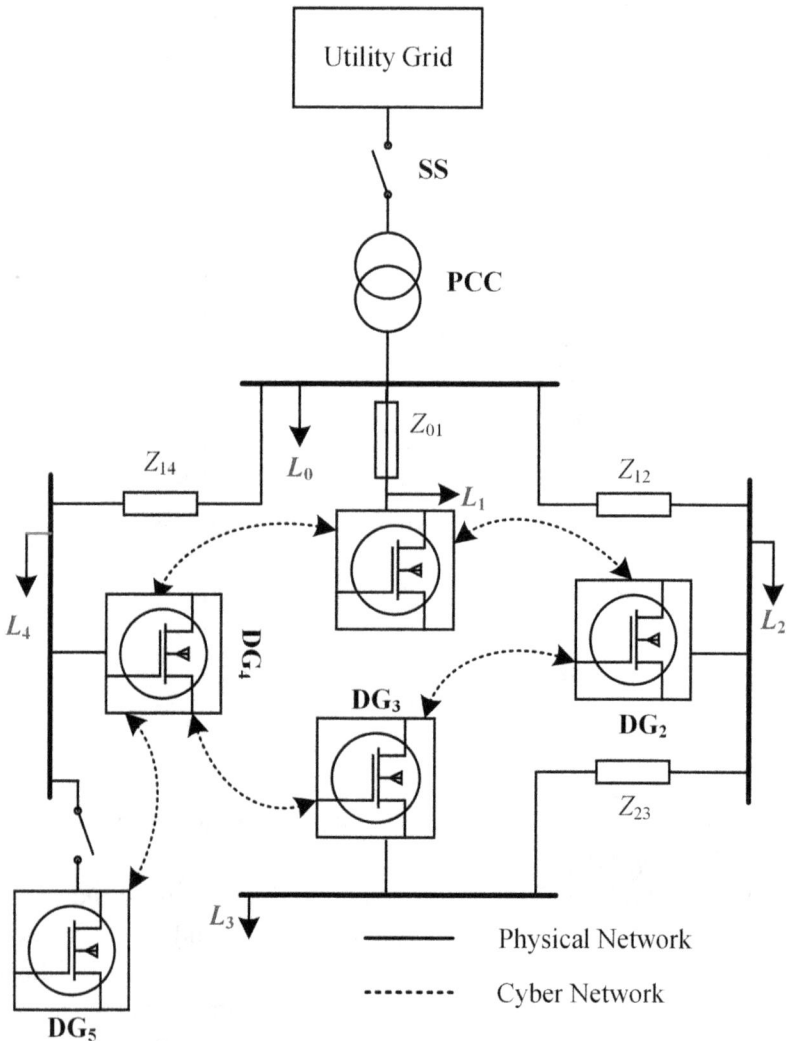

FIGURE 5.5 MG system under consideration.

rating used in this simulation study are presented in Table 5.4. Furthermore, as this study is based on distributed architecture, therefore, every DG is linked with its neighbouring DGs through CN, which adjacent matrix $A = B = [a_{cd}] = [b_{cd}]$ can be presented as:

$$A = B = \begin{pmatrix} 0 & 1 & 0 & 1 \\ 1 & 0 & 1 & 0 \\ 0 & 1 & 0 & 1 \\ 1 & 0 & 1 & 0 \end{pmatrix} \quad (5.50)$$

5.9.1 Performance of Proposed Secondary Controller

The performance of the proposed distributed synchronization controller is validated by simulating it in cases when the controller lacks synchronization capability and has the synchronization capability as presented in Figure 5.6a and b. From Figure 5.6a, it can be seen that when the SC did not receive the correction signal from the SYNC, it was unable to synchronize the VP with the UG. On the contrary, when the SC receive the signals from the SYNC then a seamless reconnection can be attained as presented in Figure 5.6b.

5.9.2 PI Controller During Synchronization

To show the performance of the distributed SC when PI controller is used to generate $\Delta\delta^s_{ref}$ and ΔE^s_{ref}, it is assumed that the MG is initially operating in an IS mode. It is considered that the MG is operated in IS mode during time interval $0 \leq t \leq 1$, and $t = 1$ sec, an Active Synchronization (AS) mode is initiated to ensure a seamless reconnection of MG with the UG. At this stage, a static switch remains in its OPEN state until a controller achieves a synchronization criteria discussed in Table 5.3. Once the criteria is met and attained for the desired time period, the command will be sent to static switch to change its state from OPEN to CLOSE and an MG is said to be in GC mode. The output waveforms of the proposed SC along with the PI controller used as SYNC, are shown in Figure 5.7. The waveforms of frequency and voltage at PCC of MG and the UG are presented in Figure 5.7a and b, respectively. Moreover, the frequency, phase angle, and voltage magnitude errors between the PCC and UG are shown in Figure 5.7c–e, respectively. From Figure 5.7c–e, it can be observed that at $t = 1$ sec, when the AS mode is initiated, the errors or differences between the waveforms are very high, but they are restored by the controller efficiently. These errors start

TABLE 5.4   Parameters Used in this Study

Parameter			Symbol	Value
Electrical				
Frequency	Switching		f_{sw}	10×10^{-3} Hz
	Nominal		f	50 Hz
	Sampling		f_a	$1/1 \times 10^{-6}$ Hz
Nominal Voltage			E	310 V
Line Impedance			Z_{01}, Z_{14}	0.8 Ω + 3 mH
			Z_{12}	1.6 Ω + 6 mH
			Z_{23}	0.9 Ω + 4 mH
	Public		L_0	2 kW + 3 kVar
Loads	Local		$L_1 = L_2 = L_3 = L_4$	1 kW + 1 kVar
LCL Filter				
Capacitance			C_f	3.31×10^{-5} F
Inductance	Inverter side		L_i	1.74×10^{-4} H
	Grid side		L_g	1.2×10^{-3} H
Static Switch Settings				
Frequency mismatch			$\lvert \Delta \omega \rvert < 5 \times 10^{-2}$ p.u.	
Phase mismatch			$\lvert \Delta \delta \rvert < 5°$	
Voltage magnitude mismatch			$\lvert \Delta E \rvert < 0.01$ p.u.	
Droop Controller				
Droop	P-ω		m_c	10^{-5} rad / (W s)
coefficients	Q-E		n_c	10^{-3} V/Var
Secondary Controller				
SC Frequency gain			$\mathbb{Z}_f$	0.1
SC Voltage gain			$\mathbb{Z}_E$	0.01
Correction Signals Generating Controllers				
PI	Proportional term	Voltage	k_p	2
	Integral term	Frequency	k_p	5
		Voltage	k_i	0.8
		Frequency	k_i	0.5
FPI	Learning parameter	β_1		2
		β_2		130
SD-based FLC	Output of GMF	b_1		2
		b_2		0.2
	Variance of GMF	σ_1		0.01
		σ_2		1
	Centre of GMF	c_1		0.9
		c_2		0.6
	Learning parameter	ς_1		0.001
		ς_2		0.025

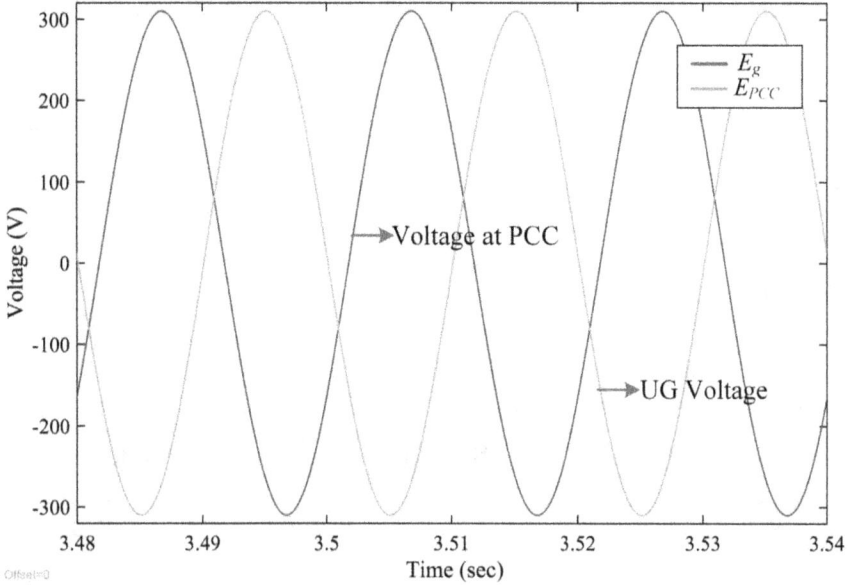

FIGURE 5.6a   Output waveforms of VP of PCC and UG without SYNC.

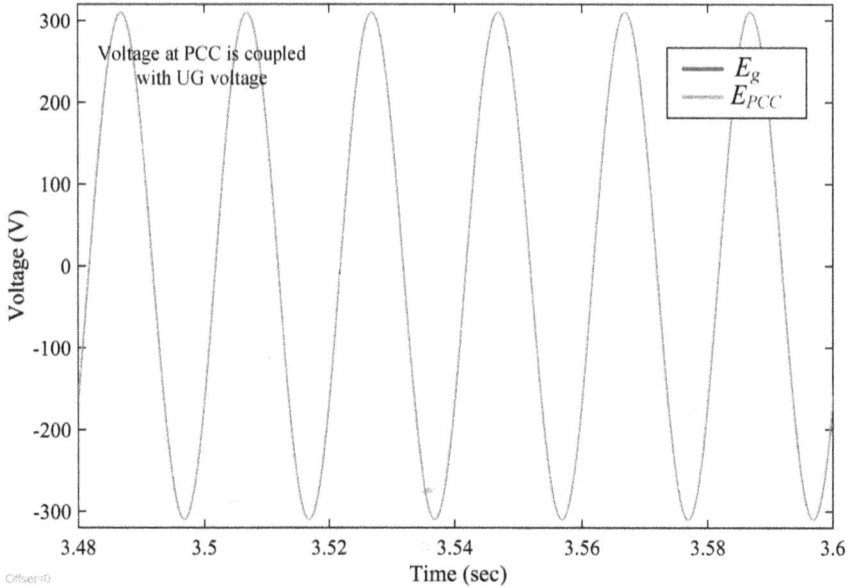

FIGURE 5.6b   Output waveforms of VP of PCC and UG with SYNC.

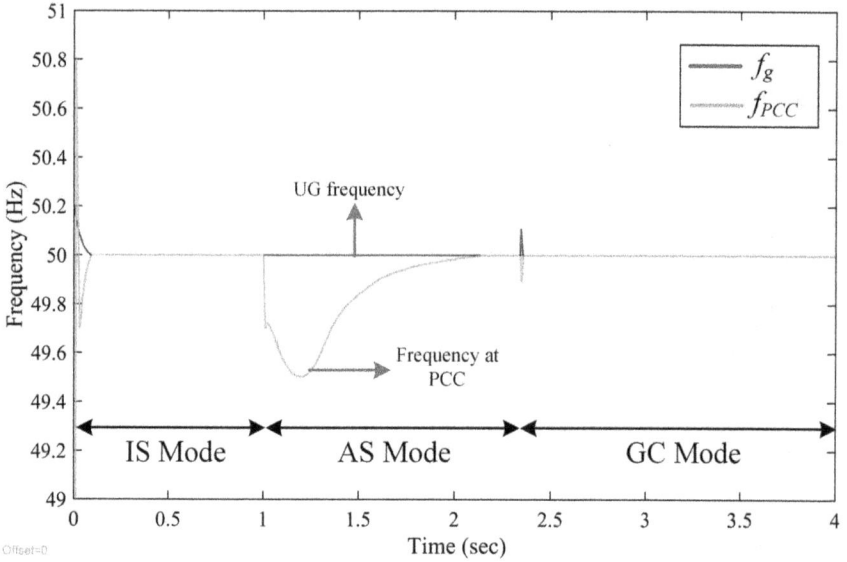

FIGURE 5.7a   Frequency at PCC and UG during synchronization by using PI controller.

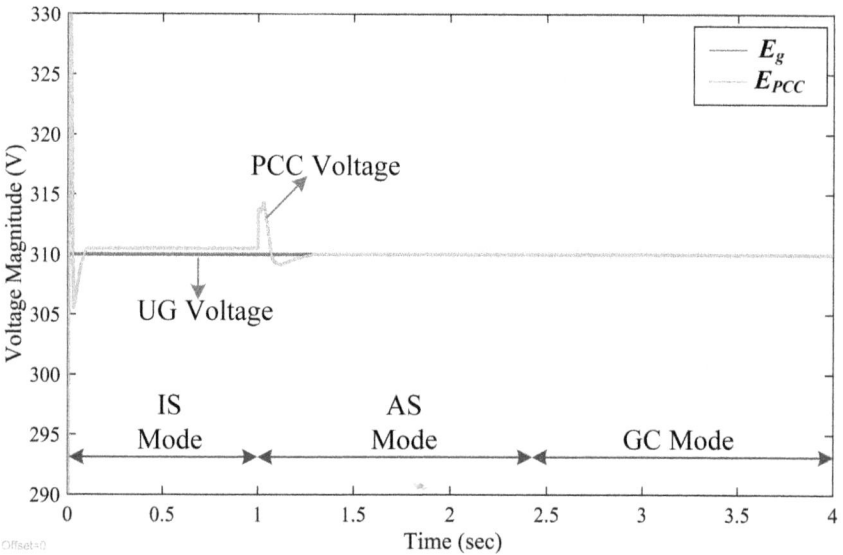

FIGURE 5.7b   Voltage magnitude at PCC and UG during synchronization by using PI controller.

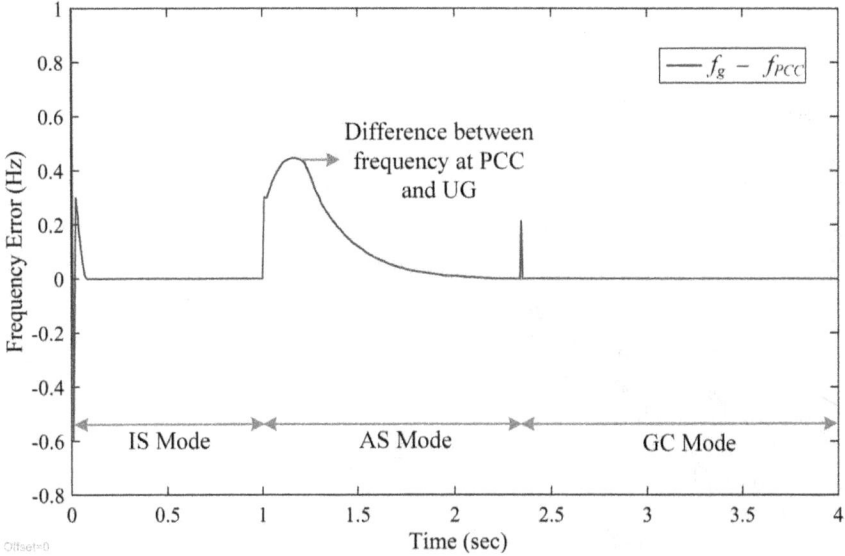

FIGURE 5.7c   Frequency difference between PCC and UG during synchronization by using PI controller.

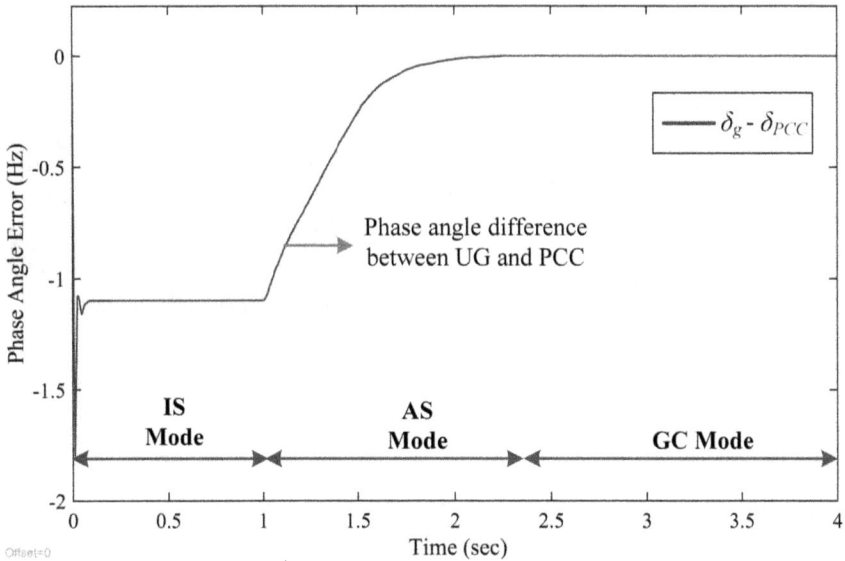

FIGURE 5.7d   Phase angle difference between PCC and UG during synchronization by using PI controller.

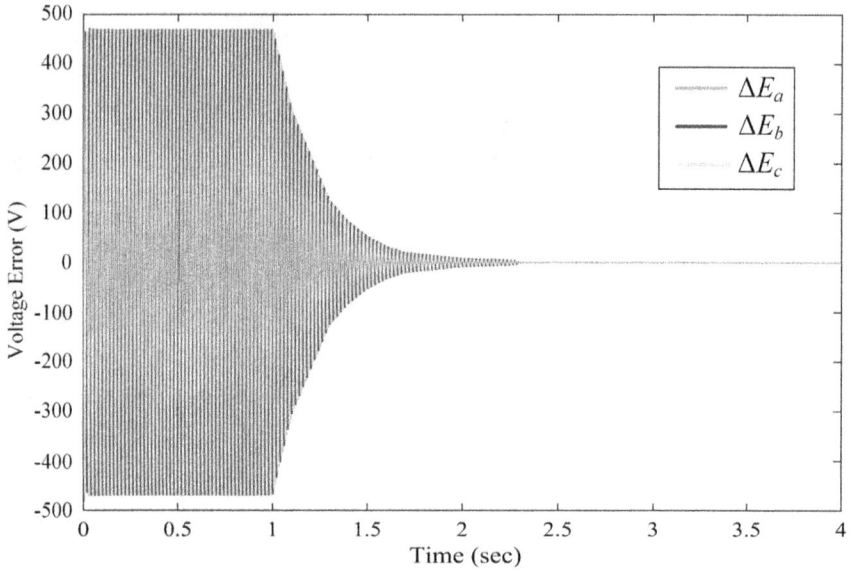

FIGURE 5.7e   Voltage magnitude error between PCC and UG during synchronization by using PI controller.

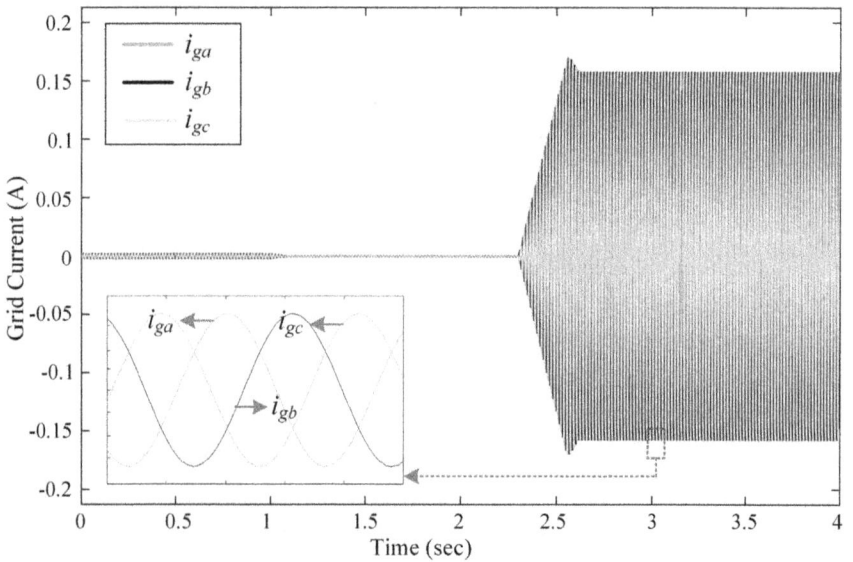

FIGURE 5.7f   UG current during synchronization by using PI controller.

decreasing gradually during the AS mode and $t=2.3$ sec; a synchronization criteria is met. Moreover, the instantaneous grid current waveforms
are presented in Figure 5.7f. From these waveforms, it can be observed
that although the synchronization criteria is met at $t=2.3$ sec, it is checked
whether the criteria will be attained for 0.1 sec or not. Hence, attaining the
criteria for more than 0.1 sec, a command to close a static switch is sent
and at $t=2.4$ sec, an MG is seamlessly reconnected with the UG.

5.9.3 FPI Controller During Synchronization

By using a PI controller as a SYNC, a seamless synchronization is attained,
but compared to high technological advancement some performance degradation (high steady-state variation and slow transient response) can be
seen in Figure 5.7. Hence, an FPI controller is used to check its performance
during synchronization, the simulation results of which are presented in
Figure 5.8. When the FPI controller is used as a SYNC, the output frequency waveforms of UG and PCC are presented in Figure 5.8a. Both the

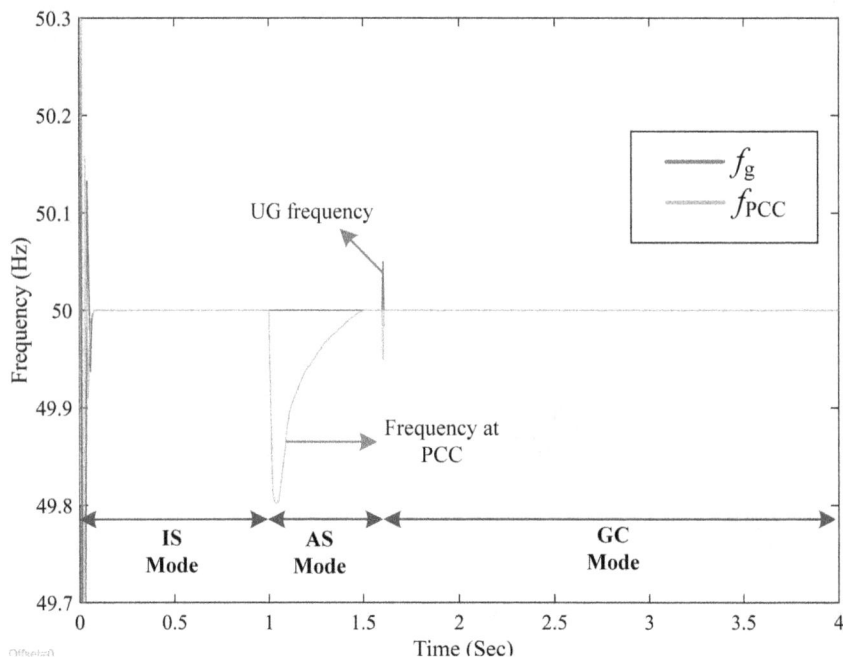

FIGURE 5.8a   Frequency at PCC and UG during synchronization by using FPI
controller.

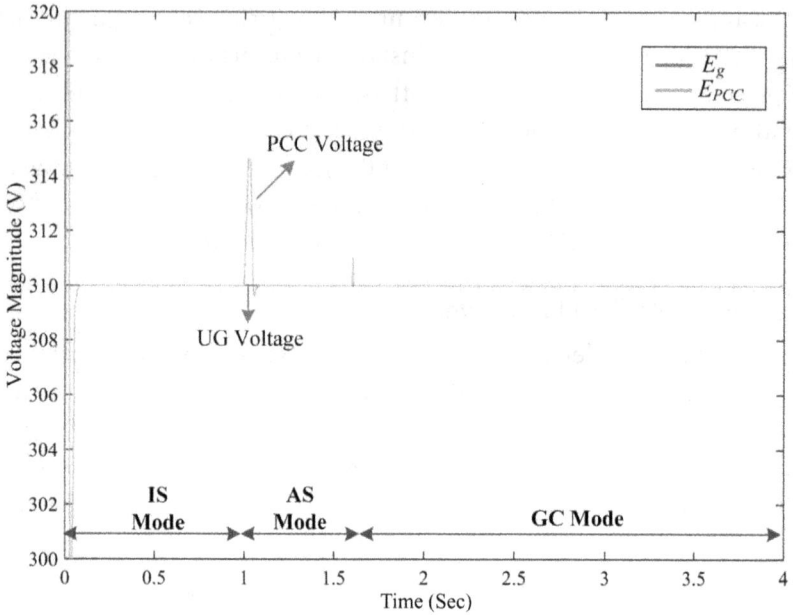

FIGURE 5.8b   Voltage magnitude at PCC and UG during synchronization by using FPI controller.

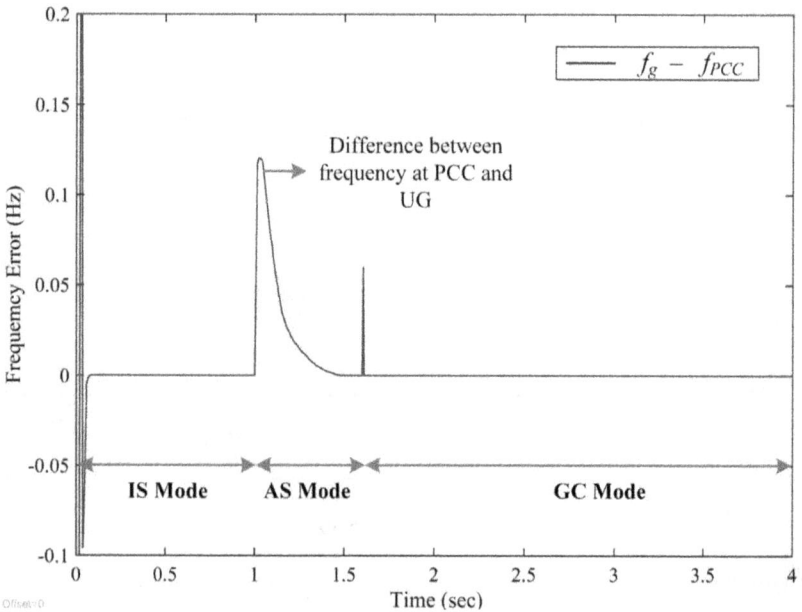

FIGURE 5.8c   Frequency difference between PCC and UG during synchronization by using FPI controller.

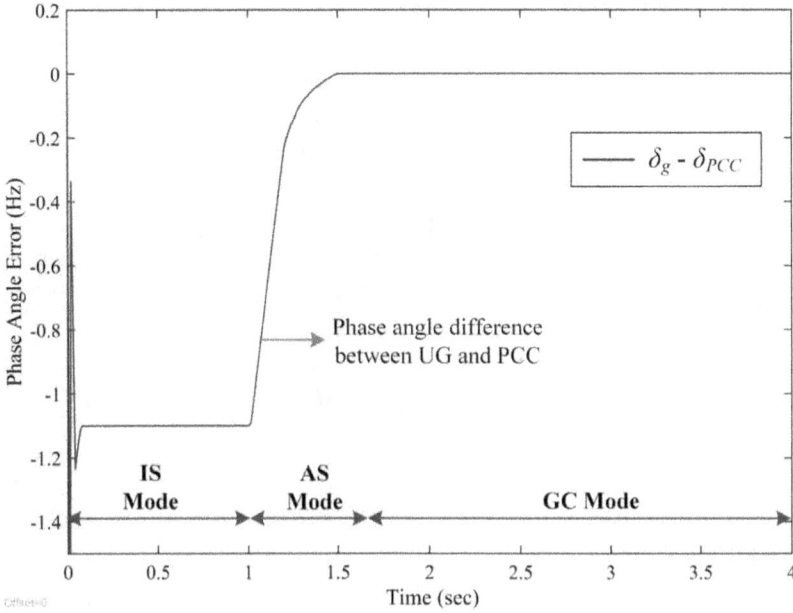

FIGURE 5.8d Phase angle difference between PCC and UG during synchronization by using FPI controller.

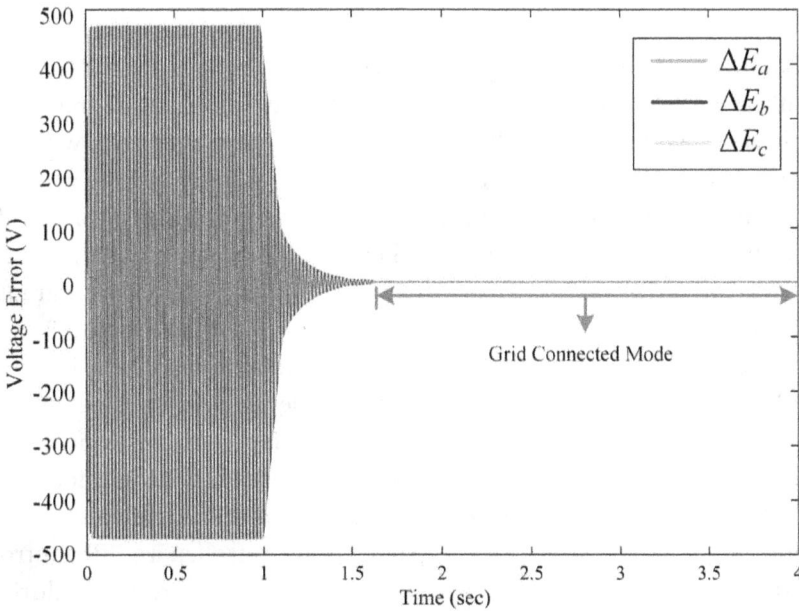

FIGURE 5.8e Voltage magnitude error between PCC and UG during synchronization by using FPI controller.

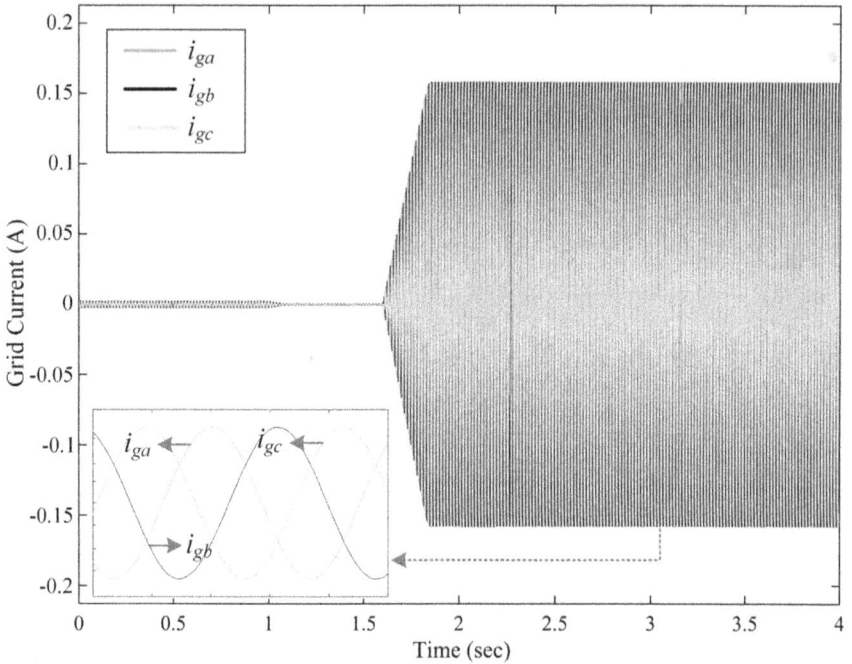

FIGURE 5.8f   UG current during synchronization by using FPI controller.

MG and the UG are operating under their nominal values until AS process is initiated. When an AS mode is activated, a deviation in MG frequency is observed while the controller restores the deviation effectively with fast transient response and the waveforms comes to its steady state at $t=1.6$ sec. Similarly, the voltage magnitude waveforms of UG and PCC of MG are presented in Figure 5.8). Moreover, when an AS mode is initiated, the frequency and phase angle errors between the PCC and UG decrease gradually and $t=1.6$ sec, the errors become zero, as presented in Figure 5.8c and d, respectively. Simultaneously, the instantaneous voltage error between the UG and PC is shown in Figure 5.8e, while the grid-injected current is presented in Figure 5.8f. From these simulation results, it can be seen that at $t=1.6$ sec, a synchronization criteria is met and is held for more than 0.1 sec; thus, a reconnection is attained at $t=1.7$ sec. Moreover, from Figure 5.8, it can be seen that compared to PI controller, an FPI controller shows small steady-state deviation and fast transient response during synchronization process.

5.9.4 SD-Based FLC During Synchronization

An SD-based FLC is used to further enhance the synchronization process by reducing the steady-state deviations and improving the transient response. The simulation results when SD-based FLC is used as the SYNC that generates the compensation signals for the SC are presented in Figure 5.9. The frequency and voltage waveforms are presented in Figure 5.9a and b, respectively, while the frequency and phase difference between UG and PCC is shown in Figure 5.9c and d, respectively. Simultaneously, the instantaneous voltage error between the UG and PC is shown in Figure 5.9e, while the grid-injected current is presented in Figure 5.9f.

From these simulation results, it can be seen that at $t=1.5$ sec, a synchronization criteria is met and is then checked whether the criteria can be held for more than 0.1 sec or not. As this controller meets the requirement of synchronization thus at $t=1.6$ sec, a seamless reconnection is achieved. Moreover, from Figure 5.9, it can be seen that compared to PI and FPI controllers, an SD-based FLC shows small steady-state deviation and fast transient response during synchronization process.

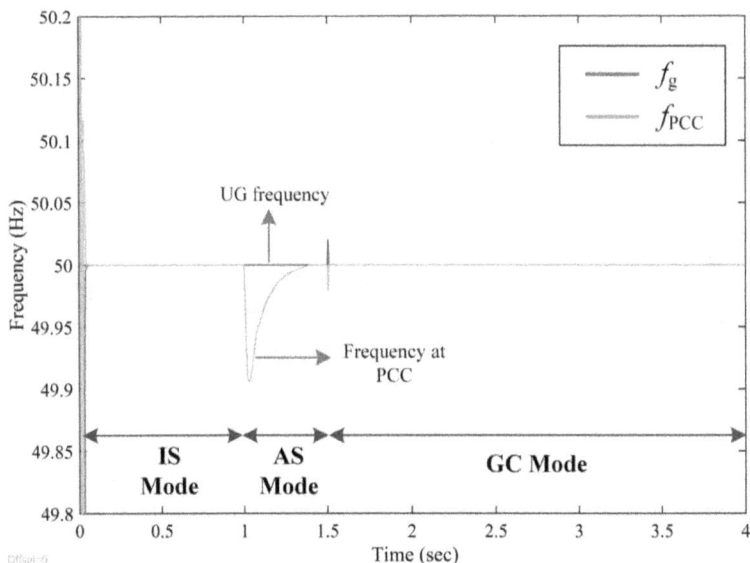

FIGURE 5.9a Frequency at PCC and UG during synchronization by using SD-based FLC controller.

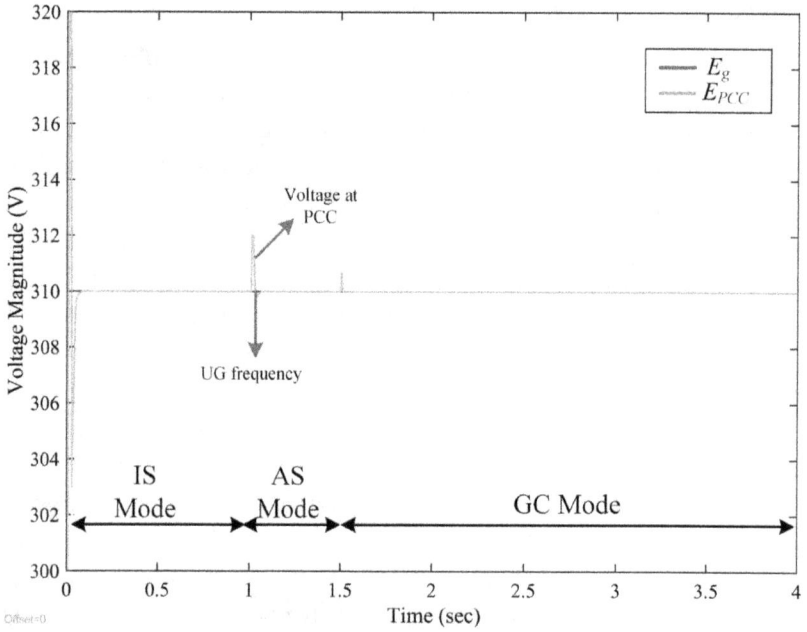

FIGURE 5.9b   Voltage magnitude at PCC and UG during synchronization by using SD-based FLC controller.

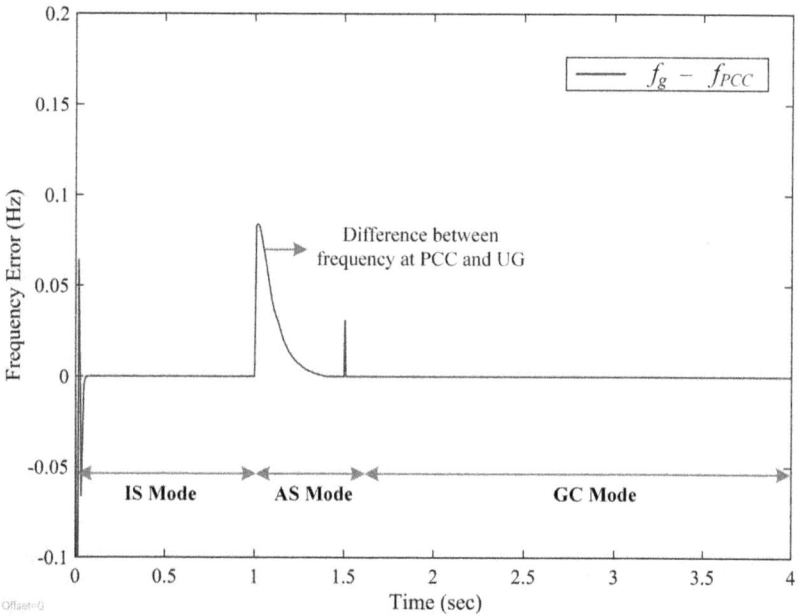

FIGURE 5.9c   Frequency difference between PCC and UG during synchronization by using SD-based FLC controller.

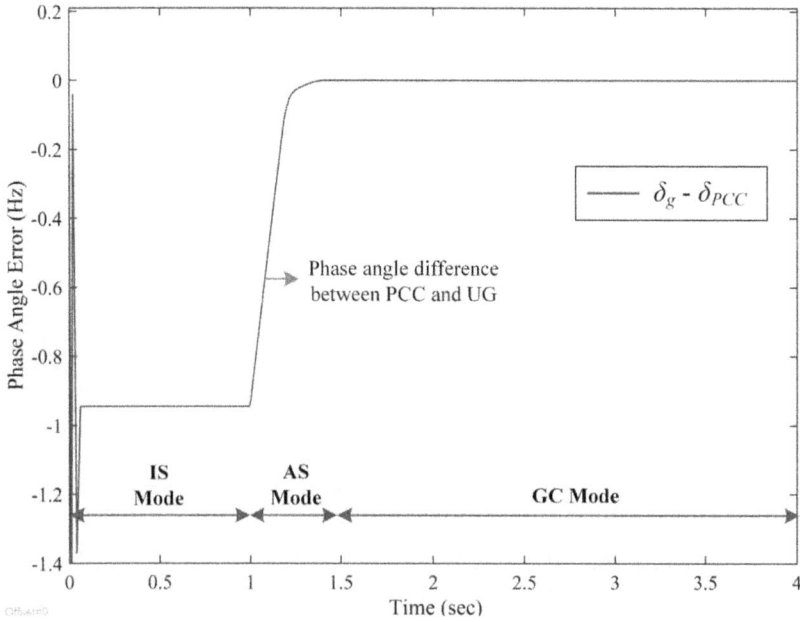

FIGURE 5.9d   Phase angle difference between PCC and UG during synchronization by using SD-based FLC controller.

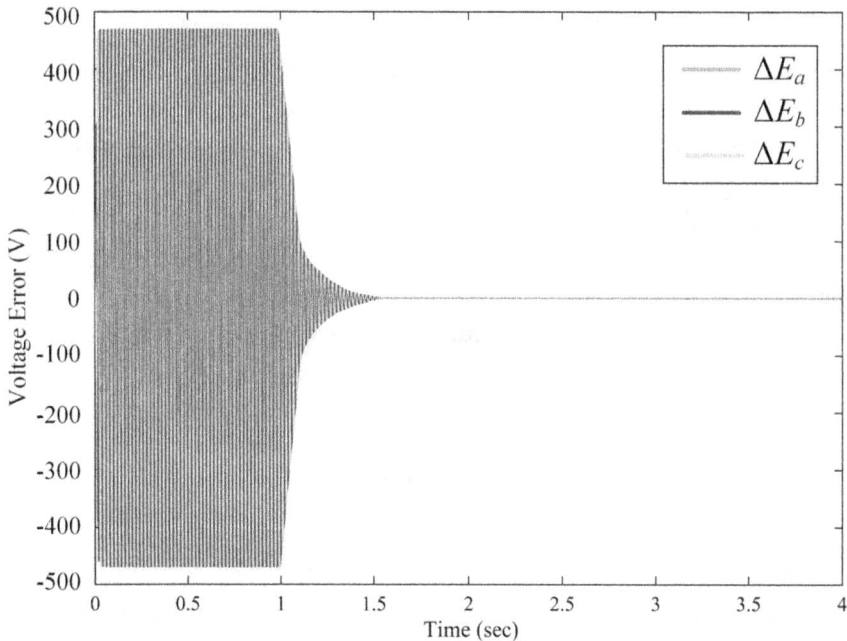

FIGURE 5.9e   Voltage magnitude error between PCC and UG during synchronization by using SD-based FLC controller.

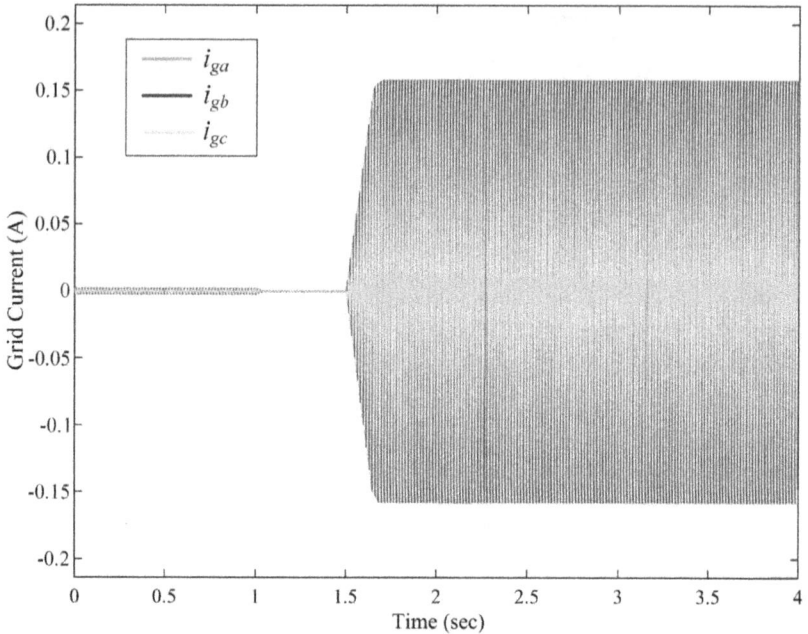

FIGURE 5.9f UG current during synchronization by using SD-based FLC controller.

5.10 CONCLUSION

In this chapter, a distributed architecture-based synchronization controller for AC MG is designed. The proposed controller ensures a seamless reconnection of IS MG with the UG. In this method, the frequency, phase angle, and voltage magnitude are controlled simultaneously and explicitly, thus resolving the natural conflict between the phase angle and frequency regulation. Moreover, an accurate power flow is achieved with this method, thus enhancing the resiliency and reliability of the system. Finally, the performance of the controller is verified through simulation results, and the results show that the proposed controller shows a fast transient response and low deviation from the references during the reconnection process. Among the proposed controllers that generate the correction signals for the SC, an SD-based FLC shows high performance as compared to PI and FPI controllers.

5.11 EXERCISES

1. In this work only one DG is considered as regulating DG. Develop a scenario in which there are two regulating DGs, and after sometime, one of the DG fails to operate; then check whether the stability of the system is ensured or not.

2. Use the controllers discussed in (4.27), (4.40), and (4.47) and put them in (5.8) to update the value of $\mathbb{Z}_f$ and use the controllers discussed in (4.58), (4.70), and (4.77) and put them in (5.10) to update the value of $\mathbb{Z}_E$. Simulate all the cases and make a comparison of their performance.

3. Consider a scenario when a synchronization controller sends the delayed signals to the secondary controller. Can an efficient performance still be achieved by the proposed controller?

REFERENCES

1. Ziouani, Islam, Djamel Boukhetala, Abdel-Moumen Darcherif, Bilal Amghar, and Ikram El Abbassi. "Hierarchical control for flexible microgrid based on three-phase voltage source inverters operated in parallel." *International Journal of Electrical Power & Energy Systems* 95 (2018): 188–201.
2. Giraldo, Jairo, Eduardo Mojica-Nava, and Nicanor Quijano. "Synchronization of isolated microgrids with a communication infrastructure using energy storage systems." *International Journal of Electrical Power & Energy Systems* 63 (2014): 71–82.
3. D'silva, Silvanus, Mohammad Shadmand, Sertac Bayhan, and Haitham Abu-Rub. "Towards grid of microgrids: Seamless transition between grid-connected and islanded modes of operation." *IEEE Open Journal of the Industrial Electronics Society* 1 (2020): 66–81.
4. Han, Hua, Xiaochao Hou, Jian Yang, Jifa Wu, Mei Su, and Josep M. Guerrero. "Review of power sharing control strategies for islanding operation of AC microgrids." *IEEE Transactions on Smart Grid* 7, no. 1 (2015): 200–215.
5. Tang, Fen, Josep M. Guerrero, Juan C. Vasquez, Dan Wu, and Lexuan Meng. "Distributed active synchronization strategy for microgrid seamless reconnection to the grid under unbalance and harmonic distortion." *IEEE Transactions on Smart Grid* 6, no. 6 (2015): 2757–2769.
6. Deng, Yan, Yong Tao, Guipeng Chen, Guangdi Li, and Xiangning He. "Enhanced power flow control for grid-connected droop-controlled inverters with improved stability." *IEEE Transactions on Industrial Electronics* 64, no. 7 (2016): 5919–5929.
7. Stringer, Norman T. "Voltage considerations during generator synchronizing." *IEEE Transactions on Industry Applications* 35, no. 3 (1999): 526–529.
8. Rizo, Mario, Francisco Huerta, Emilio Bueno, and Marco Liserre. "A synchronization technique for microgrid reclosing after islanding operation." In *IECON 2012–38th Annual Conference on IEEE Industrial Electronics Society*, pp. 5596–5601. IEEE, Montreal, QC, Canada, 2012.
9. Zhong, Qing-Chang, Phi-Long Nguyen, Zhenyu Ma, and Wanxing Sheng. "Self-synchronized synchronverters: Inverters without a dedicated synchronization unit." *IEEE Transactions on Power Electronics* 29, no. 2 (2013): 617–630.

10. Jin, Cheng, Mingzhi Gao, Xiaofeng Lv, and Min Chen. "A seamless transfer strategy of islanded and grid-connected mode switching for microgrid based on droop control." In *2012 IEEE Energy Conversion Congress and Exposition (ECCE)*, pp. 969–973. IEEE, Raleigh, NC, USA, 2012.

11. Papadimitriou, Christina N., Vasilis A. Kleftakis, and Nikos D. Hatziargyriou. "Control strategy for seamless transition from islanded to interconnected operation mode of microgrids." *Journal of Modern Power Systems and Clean Energy* 5, no. 2 (2017): 169–176.

12. Elnady, Amr, and Mohammad AlShabi. "Advanced exponential sliding mode control for microgrid at autonomous and grid-connected modes." *Bulletin of Electrical Engineering and Informatics* 10, no. 1 (2021): 474–486.

13. Sun, Chu, Geza Joos, Syed Qaseem Ali, Jean Nicolas Paquin, Carlos Mauricio Rangel, Fares Al Jajeh, Ilja Novickij, and Francois Bouffard. "Design and real-time implementation of a centralized microgrid control system with rule-based dispatch and seamless transition function." *IEEE Transactions on Industry Applications* 56, no. 3 (2020): 3168–3177.

14. Nguyen, Thai-Thanh, Hyeong-Jun Yoo, Hak-Man Kim, and Huy Nguyen-Duc. "Direct phase angle and voltage amplitude model predictive control of a power converter for microgrid applications." *Energies* 11, no. 9 (2018): 2254.

15. Awal, M. A., Hui Yu, Hao Tu, Srdjan Miodrag Lukic, and Iqbal Husain. "Hierarchical control for virtual oscillator based grid-connected and islanded microgrids." *IEEE Transactions on Power Electronics* 35, no. 1 (2019): 988–1001.

16. Nasirian, Vahidreza, Qobad Shafiee, Josep M. Guerrero, Frank L. Lewis, and Ali Davoudi. "Droop-free distributed control for AC microgrids." *IEEE Transactions on Power Electronics* 31, no. 2 (2015): 1600–1617.

17. Hou, Xiaochao, Hua Han, Chaolu Zhong, Wenbin Yuan, Yao Sun, and Mei Su. "A unified distributed control for grid-connected and islanded modes in multi-bus AC microgrid." In *IECON 2017–43rd Annual Conference of the IEEE Industrial Electronics Society*, pp. 2377–2382. IEEE, Beijing, China, 2017.

18. Hou, Xiaochao, Yao Sun, Jinghang Lu, Xin Zhang, Leong Hai Koh, Mei Su, and Josep M. Guerrero. "Distributed hierarchical control of AC microgrid operating in grid-connected, islanded and their transition modes." *IEEE Access* 6 (2018): 77388–77401.

19. Sun, Yao, Chaolu Zhong, Xiaochao Hou, Jian Yang, Hua Han, and Josep M. Guerrero. "Distributed cooperative synchronization strategy for multi-bus microgrids." *International Journal of Electrical Power & Energy Systems* 86 (2017): 18–28.

20. Du, Yuhua, Hao Tu, and Srdjan Lukic. "Distributed control strategy to achieve synchronized operation of an islanded MG." *IEEE Transactions on Smart Grid* 10, no. 4 (2018): 4487–4496.

21. Du, Yuhua, Hao Tu, Srdjan Lukic, Abhishek Dubey, and Gabor Karsai. "Distributed microgrid synchronization strategy using a novel information architecture platform." In *2018 IEEE Energy Conversion Congress and Exposition (ECCE)*, pp. 2060–2066. IEEE, Portland, OR, USA, 2018.

22. Khan, Muhammad Yasir Ali, Haoming Liu, Jie Shang, and Jian Wang. "Distributed hierarchal control strategy for multi-bus AC microgrid to achieve seamless synchronization." *Electric Power Systems Research* 214 (2023): 108910.

23. Khan, Muhammad Yasir Ali, Haoming Liu, Ren Zhang, Qi Guo, Haiqing Cai, and Libin Huang. "A unified distributed hierarchal control of a microgrid operating in islanded and grid connected modes." *IET Renewable Power Generation* 17, no. 10 (2023): 2489–2511.

24. Machowski, Jan, Zbigniew Lubosny, Janusz W. Bialek, and James R. Bumby. *Power System Dynamics: Stability and Control.* Hoboken, NJ: John Wiley & Sons, 2020.

25. Ren, Wei, and Randal W. Beard. *Distributed Consensus in Multi-Vehicle Cooperative Control.* Vol. 27, no. 2. London: Springer, 2008.

26. Simpson-Porco, John W., Qobad Shafiee, Florian Dörfler, Juan C. Vasquez, Josep M. Guerrero, and Francesco Bullo. "Secondary frequency and voltage control of islanded microgrids via distributed averaging." *IEEE Transactions on Industrial Electronics* 62, no. 11 (2015): 7025–7038.

27. Shi, Di, Xi Chen, Zhiwei Wang, Xiaohu Zhang, Zhe Yu, Xinan Wang, and Desong Bian. "A distributed cooperative control framework for synchronized reconnection of a multi-bus microgrid." *IEEE Transactions on Smart Grid* 9, no. 6 (2017): 6646–6655.

28. Mohamed, Yasser Abdel-Rady Ibrahim, and Ehab F. El-Saadany. "Adaptive decentralized droop controller to preserve power sharing stability of paralleled inverters in distributed generation microgrids." *IEEE Transactions on Power Electronics* 23, no. 6 (2008): 2806–2816.

29. IEEE Standards Board. *IEEE Standard for Interconnecting Distributed Resources with Electric Power Systems: 1547-2003.* Piscataway, NJ: IEEE, 2003.

30. Rafi, Fida Hasan Md, Md Jahangir Hossain, Md Shamiur Rahman, and Seyedfoad Taghizadeh. "An overview of unbalance compensation techniques using power electronic converters for active distribution systems with renewable generation." *Renewable and Sustainable Energy Reviews* 125 (2020): 109812.

31. Wilamowski, Bogdan M., and Hao Yu. "Improved computation for Levenberg–Marquardt training." *IEEE Transactions on Neural Networks* 21, no. 6 (2010): 930–937.

CHAPTER 6

Secondary Control Strategy for Microgrid Considering Communication Delays

6.1 INTRODUCTION AND LITERATURE REVIEW

In a distributed control architecture, the exchange of information between the DGs is usually performed through CN. However, one of the main limitations that affects the transient performance and stability of the system during the exchange of information is the CTDs, which are usually caused by the limited CN bandwidth or network traffic. In case of any CTDs in a CN, a DG will receive delayed information from its neighbouring DGs and from their references that can affect the efficacy of the controller's performance [1,2].

Different controllers and methods are designed by the researchers to cope with the effects of CTDs on the MG system's performance. For instance, the authors in Refs. [3,4] proposed a SC to mitigate the effects of CTDs on the system's performance. In these works, the authors only consider the effects of fixed CTDs on the system's performance. Likely, the authors in Ref. [5] proposed a stochastic consensus-based SC to mitigate the effects of CTDs on frequency and voltage regulation. In these works [3–5], it is considered that there are uniform and fixed delays in all the cyber-links;

 DOI: 10.1201/9781003594284-6

however, in practical systems, the CTDs are varying rather than uniform. Hence, to overcome this drawback, a master-slave architecture-based distributed SC is designed in Ref. [6,7] that considers non-uniform CTDs in the cyber-links. However, due to master-slave architecture, a new uncertainty that may occur is that the failure of leader DG may cause system instability [8]. The authors in Ref. [9] presented a finite-time distributed control architecture that uses an adjustable graph gain to mitigate the effects of CTDs on frequency regulation and active power sharing. In this method, it is mandatory to optimize the CN topology in advance; moreover, the other limitation of this work is that the authors only consider small CTDs. Furthermore, the authors consider numerous leader DGs that may cause an increase in computational complexity.

A model predictive-based SC for frequency regulation and active power sharing in the presence of CTDs is proposed in Ref. [10]. This method ensures fast consensus convergence and significantly enhances the system's performance. However, this research lacks a discussion of the design of voltage regulation and reactive power sharing. A distributed SC based on FLC is presented in Ref. [11] to cope with the effects of CTDs. In this controller, the gain coefficients of the cooperative SC are optimized by using FLC to improve the output voltage waveforms of the MG system under the occurrence of CTDs. On the contrary, this method involves the manual extraction of fuzzy features of FLC that make it less robust and flexible for different disturbances [12].

An H∞-based SC for an IS MG considering CTDs is proposed in Ref. [13]. In this controller, the main consideration is given to the upper bound of delay, while no consideration is given to the enhancement of the CTD margin. In Ref. [14], an average prediction method-based distributed SC to mitigate the effect of CTDs in IS MG is presented. Linear matrix inequalities and Lyapunov function are used to study the CTDs. Furthermore, a delay margin is significantly enhanced in this method; however, in case of multiple CTDs, high-performance degradation of the controller is observed. A two-layer distributed controller to mitigate the effects of CTDs is proposed in Ref. [15]. This controller significantly improves the performance of an MG system in the presence of varying multiple CTDs but its dual-layered structure might increase its implementation complexity. Moreover, it is specifically designed for the DC MG system; its implementation in an AC MG system needs further improvements.

An SC based on an event-triggered method is designed for an MG considering CTDs in Ref. [16]. In this controller, it is considered that all the cyber-links between the DGs have the same CTDs; however, this is practically not feasible, as in large MG systems, DGs are far away from each other therefore face different CTDs. Furthermore, the authors in this work only consider the effects of CTDs on voltage regulation while no discussion is provided regarding the impacts of CTDs on frequency regulation. Similarly, the authors in Ref. [17] proposed an average consensus-based delay-tolerant SC. In this method, the authors consider that all the cyber-links have the same CTDs; moreover, the mathematical model that validates the stability of the system is also not discussed. The authors in Ref. [18] proposed an event-triggered-based SC that efficiently mitigates the effect of varying CTDs on the system's performance. A Lyapunov function is used to ensure system stability and analyse the triggering conditions. Although this controller performs efficiently, this work lacks an explanation of the design of the controller's parameters. Another SC based on an event-triggered mechanism presented in Ref. [19] considers non-uniform CTDs. In this technique, although the communication burden is reduced significantly, it lacks to discuss any explanation about the system's stability. Furthermore, a comparative analysis of different control techniques along with the proposed controller is presented in Table 6.1.

TABLE 6.1 Comparative Analysis Different Controller

Ref	DD	VD	FD	BU	APF	ASCGP	PnP
[3]	√	×	√	×	NG	×	×
[4]	×	×	√	√	√	×	√
[6]	×	√	√	√	√	×	√
[9]	√	√	√	√	√	×	√
[10]	×	√	√	×	NG	×	×
[11]	√	×	√	×	√	√	√
[13]	√	√	√	√	√	×	√
[14]	√	√	√	√	√	×	×
[17]	√	×	√	√	√	×	√
[18]	√	√	√	√	√	×	√
P	×	√	√	×	√	√	√

Abbreviations: √, Presence of Feature; ×, Absence of Feature; APF, Appropriate Power Flow; ASCGP, Adaptive Secondary Control Gain Parameters; BU, Buffer Used; DD, Delay-Dependent; FD, Fixed Delays; H, High; M, Medium; NG, Not Given; P, Proposed; VD, Variable Delays.

6.2 CONTRIBUTIONS

The state-of-the-art literature discussed above mainly focuses on frequency restoration, system stability, optimal power sharing, and voltage regulation in the presence of CTDs. In addition to the numerous advantages of the above-discussed works, there are some limitations that need to be considered. Some of the works consider uniform and small CTDs in all the cyber-links; some of the works present controllers that are delay-dependent, and some of the works use a compensator to mitigate the effects of CTDs. Hence, to cope with these limitations and challenges, this chapter presents a delay-independent DAPI controller that having the following main features:

- This controller can mitigate the effects of uniform and non-uniform as well as small and large CTDs.

- As most of the controllers are delay-dependent, therefore, a compensator is used in those controllers to minimize the CTD effects. However, the compensator requires storage, power, and computation which is not desirable. Moreover, the memory of the compensator becomes limited when the number of participants increases; hence, it may not be a feasible solution [20]. Therefore, a presented controller avoids the use of any compensator or buffer.

- The presented controller is implemented in a distributed architecture, thus enabling the PnP functionality. Moreover, it shows high efficacy and robustness against varying CTDs and sudden load variation conditions.

6.3 SYSTEM CONFIGURATION

In this section, a detailed description of the physical-cyber network used in this work is provided.

6.3.1 Cyber System

A CN is an MG system that can be modelled by using a diagraph that can be presented as $G = (N, \xi)$; where a set of nodes (DGs) is presented by N such that $N = \{1,...n\}$; a set of edges (i.e. links) is denoted by ξ, i.e., $\xi = \{e_1,...,e_m\}$ such that $\xi = [N]^2$. An adjacent matrix of G can be presented as $A = R^{|N||N|}$ that consists of elements a_{cd}, where R presents a set of real numbers. If there is a link or edge between two neighbouring nodes, i.e., nodes c and d, then $a_{cd} = a_{dc} = 1$, else $a_{cd} = a_{dc} = 0$. The degree of c_{th}-node can be presented as $D_c = \displaystyle\sum_{c,d \in |N|} a_{cd}$, where $c \neq d$. The Laplacian matrix

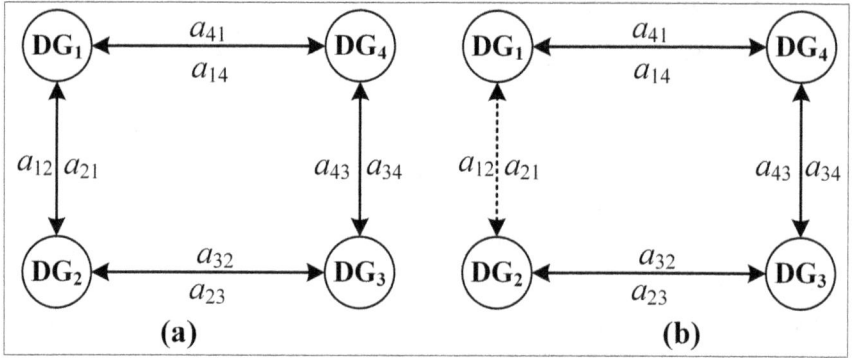

FIGURE 6.1   Bidirectional connected ring cyber network.

(ℓ) of the diagraph can be given as $\ell = D - A$, where $D = \mathrm{diag}(D_c) \in R^{|N||N|}$ represents the degree of the matrix. Moreover, the nodes are arranged in manner that if there is an edge or path between nodes c and d then the graph is said to be connected. Hence, if G is connected, then ℓ is a positive semi-definite matrix having an eigen-value of zero and have an eigen-vector 1_n, such that $\ell 1_n = 0_n$ [21].

In prospect of this research, in a diagraph, the nodes present the DGs and the edges represents the cyber-links between the DGs; a schematic of which is presented in Figure 6.1a that considers no CTDs. Hence, ℓ along with D and A can be presented as:

$$A = \begin{bmatrix} 0 & 1 & 0 & 1 \\ 1 & 0 & 1 & 0 \\ 0 & 1 & 0 & 1 \\ 1 & 0 & 1 & 0 \end{bmatrix} \tag{6.1a}$$

$$D = \begin{bmatrix} 2 & 0 & 0 & 0 \\ 0 & 2 & 0 & 0 \\ 0 & 0 & 2 & 0 \\ 0 & 0 & 0 & 2 \end{bmatrix} \tag{6.1b}$$

$$L = \begin{bmatrix} 2 & -1 & 0 & -1 \\ -1 & 2 & -1 & 0 \\ 0 & -1 & 2 & -1 \\ -1 & 0 & -1 & 2 \end{bmatrix} \tag{6.1c}$$

A CN shown in Figure 6.1a considers that there are no CTDs in the CN; however, in practical networks, the CTDs are inherent in the CN. Therefore, in case of any CTDs, a DG will receive delayed data from its adjacent DGs that may affect the performance of the system. Assume that $h_{cd} > 0$ is the CTD between DG_1 and DG_2. Then, for easy understanding, the cyber-link between them is marked with a dotted line, as presented in Figure 6.1b. If the controller is unable to compensate the CTDs, then the system stability can be affected by the delayed information. Hence, it is very important to study the effects of CTDs

6.3.2 Physical System

A generalized meshed MG system with n number of DGs is considered to have an Admittance (Y_{cd}) between c_{th}-DG and d_{th}-DG and can be presented as $Y_{cd} = G_{cd} + jB_{cd} \in C$, in which $G_{cd} \in R$ and $B_{cd} \in R$ denotes the conductance and susceptance, respectively. In a scenario where there is no link among c_{th}-DG and d_{th}-DG, then the admittance can be given as $Y_{cd} = 0$. On the contrary, when there is the connection of c_{th}-DG with its neighbours, then, a set of neighbouring DGs can be presented as $N_c = \{d | d \in N, c \neq d, Y_{cd} \neq 0\}$. Moreover, it is considered that the MG system is strongly connected then for all the pairs $\{c,d\} \in N \times N, c \neq d$ in an MG there is a sequence of nodes from c and d such that any two consecutive connected nodes there exists a power line that shows the admittance [22].

Hence, on the basis of these considerations, the active and reactive power injected by c_{th}-DG can be presented as Ref. [23]:

$$P_c = E_c^2 G_{cc} - \sum_{d=1}^{n} E_c E_d |Y_{cd}| \cos(\delta_c - \delta_d - \phi_{cd}) \qquad (6.2)$$

$$Q_c = -\left(\sum_{d=1}^{n} E_c E_d |Y_{cd}| \sin(\delta_c - \delta_d - \phi_{cd}) + E_c^2 B_{cc} \right) \qquad (6.3)$$

where $|Y_{cd}| = \sqrt{G_{cd}^2 + B_{cd}^2}$ and the angle of admittance is denoted by ϕ_{cd}. Furthermore, it is considered that the transmission lines are purely inductive and lossless. Thus, $G_{cd} = 0$, $Y_{cd} = jB_{cd}$, and $\phi_{cd} = \phi_{dc} = -(\pi/2)$. Hence, (6.2) and (6.3) on the basis of these considerations can be given as:

$$P_c = \sum_{d=1}^{n} E_c E_d |B_{cd}| \cos(\delta_c - \delta_d) \qquad (6.4)$$

$$Q_c = E_c^2 \sum_{d=1}^{n} B_{cd} - \sum_{d=1}^{n} E_c E_d |B_{cd}| \cos(\delta_c - \delta_d) \qquad (6.5)$$

6.4 DISTRIBUTED FREQUENCY CONTROLLER

A delay-independent distributed secondary frequency controller that mitigate the effects of CTDs is given as:

$$\omega_c = \omega^* - m_c P_c + \varpi_c^\omega \qquad (6.6)$$

$$\frac{d\varpi_c^\omega}{dt} = F_f \left[\sum_{d=1}^{n} a_{cd} \left(\omega_d(t-\tau) - \omega_c(t) \right) - \left(\omega_c - \omega^* \right) \right] \qquad (6.7)$$

where ω_c and ω_d are the angular frequency of c_{th}-DG and d_{th}-DG, respectively $(c,d = 1,2,\ldots n)$; ω^* is the reference or nominal angular frequency; m_c is the $(P_c - \omega_c)$ droop coefficient and is given as $m_c = (\omega_{max} - \omega_{min})/P_{max}$; ϖ_c^ω is the SC variable; F_f is the gain of the SC; if there exists a cyber-link among c_{th}-DG and d_{th}-DG then $a_{cd} = 1$, while when there is no link then $a_{cd} = 0$; a CTD is presented by $\tau(t)$ and $\tau(t) \in [0,h]$ is a bounded time-varying delay with $h > 0$.

An expression in (6.7) demonstrates that the controller is capable of regulating the frequencies of the DGs to its nominal values in the presence of CTDs. To present the stability condition, define an error function as:

$$e_c^\omega(t) = \omega_c(t) - \omega^* \qquad (6.8)$$

As discussed above, it is considered that all the DGs are connected through a sparse CN and have access to their nominal values, therefore, they must be synchronized with their references even if there exist some CTDs. Hence, it can be said that $\dot{\omega}^* = 0$. By taking the time derivative of (6.8) and considering the assumption $\dot{\omega}^* = 0$, we get:

$$\dot{e}_c^\omega(t) = \dot{\omega}_c(t) \qquad (6.9)$$

Put (6.9) in (6.7) would yield us to:

$$\frac{d\varpi_c^\omega}{dt} = F_f \left[\sum_{d=1}^{n} a_{cd} \left(e_d^\omega(t-\tau) - e_c^\omega(t) \right) - \left(e_c^\omega(t) \right) \right] \qquad (6.10)$$

To derive the conditions for the system stability, a Lyapunov-Krasovskii function for (6.9) can be expressed as:

$$V_f(t) = \sum_{c=1}^{n} \sum_{d=1}^{n} a_{cd} \int_{t-\tau}^{t} \left(e_d^{\omega}(s) \right)^2 ds + \sum_{c=1}^{n} \left(e_c^{\omega}(t) \right)^2 \qquad (6.11)$$

From (6.11), it is obvious that $V_f(t) \geq 0$; and $V_f(t) = 0$ if and only if $e_c^{\omega}(t) = e_d^{\omega}(t) = 0$, that is $\omega_c(t) = \omega_d(t) = \omega^*$. Take a derivative of this Lyapunov function would give us:

$$\dot{V}_f(t) = \sum_{c=1}^{n} \sum_{d=1}^{n} a_{cd} \left(\left(e_d^{\omega}(t) \right)^2 - \left(e_d^{\omega}(t-\tau) \right)^2 \right) + 2 \sum_{c=1}^{n} e_c^{\omega}(t) \dot{e}_c^{\omega}(t) \qquad (6.12)$$

Solving the above expression would yield us to:

$$\dot{V}_f(t) = 2 \left[\sum_{c=1}^{n} \sum_{d=1}^{n} a_{cd} e_c^{\omega}(t) \left(e_d^{\omega}(t-\tau) - e_c^{\omega}(t) \right) - \sum_{c=1}^{n} \left(e_c^{\omega}(t) \right)^2 \right]$$

$$+ \sum_{c=1}^{n} \sum_{d=1}^{n} a_{cd} \left(\left(e_d^{\omega}(t) \right)^2 - \left(e_d^{\omega}(t-\tau) \right)^2 \right) \qquad (6.13)$$

As the diagraph is balance, hence:

$$\sum_{c=1}^{n} \sum_{d=1}^{n} a_{cd} \left(e_c^{\omega}(t) \right)^2 = \sum_{c=1}^{n} \sum_{d=1}^{n} a_{cd} \left(e_d^{\omega}(t) \right)^2 \qquad (6.14)$$

Apply (6.14) in (6.13) would yields us to:

$$\dot{V}_f(t) = - \left[2 \sum_{c=1}^{n} \left(e_c^{\omega}(t) \right)^2 + \sum_{c=1}^{n} \sum_{d=1}^{n} a_{cd} \left(e_c^{\omega}(t) - e_d^{\omega}(t-\tau) \right)^2 \right] \qquad (6.15)$$

From (6.15), it can be concluded that $\dot{V}_f(t) < 0$ or $\dot{V}_f(t) = 0$. When $\dot{V}_f(t) < 0$, it means that DGs' frequencies are converging to the reference value (i.e. ω^*) asymptotically. When $\dot{V}_f(t) = 0$ then we get $e_k^{\omega}(t) = 0$ and $e_c^{\omega}(t) = e_d^{\omega}(t-\tau)$.

Remark 6.1

At steady-state, a SC input ϖ_c^ω is activated and can be calculated as $\varpi_c^\omega = \int_{t_0}^{t} F_f \left[\sum_{d=1}^{n} a_{cd} \left(\omega_d(t-\tau) - \omega_c(t) \right) - \left(\omega_c(t) - \omega^* \right) \right] dt.$
At a new steady-state, i.e., $\omega_c = \omega^*$, then in light of (6.6), we have

$$\omega_c - \omega^* + m_c P_c = \int_{t_0}^{t} F_f \left[\sum_{d=1}^{n} a_{cd} \left(\omega_d(t-\tau) - \omega_c(t) \right) - \left(\omega_c(t) - \omega^* \right) \right] dt.$$

Thus, it can be said that ϖ_c^ω is added to restore the frequency of the system to its nominal reference value.

6.5 ACTIVE POWER-SHARING CONTROLLER

A delay-independent active power-sharing controller can be presented as:

$$\omega_c = \omega^* - m_c P_c + \varpi_c^P \tag{6.16}$$

$$\frac{d\varpi_c^P}{dt} = P_P \left[\sum_{d=1}^{n} a_{cd} \left(\varpi_d^P(t-\tau) - \varpi_c^P(t) \right) \right] \tag{6.17}$$

where ϖ_c^P is the SC variable; P_P is the gain of the SC; and τ is the CTD between c_{th}-DG and d_{th}-DG. For simplicity, the dynamics in (6.17) can be presented in vector forms as:

$$\frac{d\varpi_c^P}{dt} = \mathbb{Z}_P \varpi(t) + \mathbb{Z}_{P-1} \varpi(t - \tau(t)) \tag{6.18}$$

where $\mathbb{Z}_P$ and $\mathbb{Z}_{P-1}$ are $n \times n$ matrices whose elements are positive real numbers. To derive the stability conditions for the system, a Razumikhin's method is used by applying a Lyapunov function as:

$$V_P(\varpi_t) = \varpi^T(t) K_P \varpi(t) \tag{6.19}$$

where K_P is $n \times n$ matrix and $K_P > 0$, it should be noted that K_P is symmetric and positive definite. Consider a time derivative of $V_P(\varpi_t)$ along (6.18) and apply the Lyapunov Razumikhin's method as $\bar{\varsigma}_P(s) = \bar{\varsigma}_P.s$, where

$\bar{\varsigma}_P > 0$. Hence, according to Razumikhin's theorem presented in Refs. [24,25] when the condition:

$$\bar{\varsigma}_P \varpi^T(t) K_P \varpi(t) - \varpi^T(t-\tau(t)) K_P \varpi(t-\tau(t)) \geq 0 \qquad (6.20)$$

holds for some constant such that $\bar{\varsigma}_P = 1 + \varsigma_P$, where $\varsigma_P > 0$. Thus, it can be said that for any $\kappa_P > 0$ there exists $\gamma_P > 0$ such that:

$$\dot{V}_P(\varpi(t)) = 2\varpi^T(t) K_P \left[\mathbb{Z}_P \varpi(t) + \mathbb{Z}_{P-1} \varpi(t-\tau(t)) \right]$$

$$\leq 2\varpi^T(t) K_P \left[\mathbb{Z}_P \varpi(t) + \mathbb{Z}_{P-1} \varpi(t-\tau(t)) \right] \qquad (6.21)$$

$$+\kappa_P \left[\bar{\varsigma}_P \varpi^T(t) K_P \varpi(t) - \varpi^T(t-\tau(t)) K_P \varpi(t-\tau(t)) \right] \leq -\gamma_P |\varpi(t)|^2$$

If

$$\begin{bmatrix} \mathbb{Z}_P^T K_P + K_P \mathbb{Z}_P + \kappa_P K_P & K_P \mathbb{Z}_{P-1} \\ \mathbb{Z}_{P-1}^T K_P & -\kappa_P K_P \end{bmatrix} < 0 \qquad (6.22)$$

From (6.21) and (6.22), it can be seen that the matrix inequality is independent from the delay derivative term; hence, it can be concluded that there is no constraint on delay derivative. Hence, (6.21) provide sufficient conditions for the asymptotically and independent stability of the system having CTDs.

6.6 DISTRIBUTED VOLTAGE CONTROLLER

A distributed secondary voltage controller that mitigates the effects of CTDs is given as:

$$E_c = E^* - n_c Q_c + v_c^E \qquad (6.23)$$

$$\frac{dv_c^E}{dt} = E_E \left[\sum_{d=1}^{n} b_{cd} \left(E_d(t-\tau) - E_c(t) \right) - \phi_c \left(E_c - E^* \right) \right] \qquad (6.24)$$

where E_c and E_d are the voltage of c_{th}-DG and d_{th}-DG, respectively $(c,d=1,2,n)$; E^* is the reference voltage; v_c^E is the SC variable; E_E and ϕ_c are the positive gains of SC; if a communication link among c_{th}-DG and d_{th}-DG exists, then $b_{cd} = 1$, else $b_{cd} = 0$. Furthermore, to avoid extra CN it is considered that $B = A \Rightarrow \{b_{kl}\} = \{a_{kl}\}$.

A main objective of the SC presented in (6.24) is to regulate the voltages of DGs to their reference values in the presence of CTDs. Hence, to derive a stability condition, we define an error function as:

$$e_c^E(t) = E_c(t) - E^*(t) \tag{6.25}$$

Let us consider that all the DGs are connected through a sparse CN and have access to their nominal values; therefore, they must be synchronized with their references even if there exist some CTDs. Hence, it can be said that $\dot{E}^*(t) = 0$. By taking the time derivative of (6.25) and considering the assumption $\dot{\omega}^* = 0$, we get:

$$\dot{e}_c^E(t) = \dot{E}_c(t) \tag{6.26}$$

Put (6.26) in (6.24) would yield us to:

$$\frac{dv_c^P}{dt} = E_E\left[\sum_{d=1}^{n} b_{cd}\left(e_d^E(t-\tau) - e_c^E(t)\right) - \phi_c\left(e_c^E(t)\right)\right] \tag{6.27}$$

To derive the conditions, a Lyapunov-Krasovskii function for (6.27) can be expressed as:

$$V_E(t) = \sum_{c=1}^{n}\sum_{d=1}^{n} b_{cd}\int_{t-\tau}^{t}\left(e_d^E(s)\right)^2 ds + \sum_{c=1}^{n}\left(e_c^E(t)\right)^2 \tag{6.28}$$

Take a derivative of this Lyapunov function along (6.27) would give us:

$$\dot{V}_E(t) = 2\left[\sum_{c=1}^{n}\sum_{d=1}^{n} b_{cd}e_c^E(t)\left(e_d^E(t-\tau) - e_c^E(t)\right) - \sum_{c=1}^{n}\phi_c\left(e_c^E(t)\right)^2\right] +$$
$$\sum_{c=1}^{n}\sum_{d=1}^{n} b_{cd}\left(\left(e_d^E(t)\right)^2 - \left(e_d^E(t-\tau)\right)^2\right) \tag{6.29}$$

As the diagraph is balance, hence:

$$\sum_{c=1}^{n}\sum_{d=1}^{n} b_{cd}\left(e_c^E(t)\right)^2 = \sum_{c=1}^{n}\sum_{d=1}^{n} b_{cd}\left(e_d^E(t)\right)^2 \tag{6.30}$$

Applying (6.30) in (6.29) would yields us to:

$$\dot{V}_E(t) = -\left[2\sum_{c=1}^{n} \phi_c \left(e_c^E(t) \right)^2 + \sum_{c=1}^{n}\sum_{d=1}^{n} b_{cd} \left(e_c^E(t) - e_d^E(t-\tau) \right)^2 \right] \quad (6.31)$$

From (6.31), it can be concluded that $\dot{V}_E(t) < 0$ or $\dot{V}_E(t) = 0$. When $\dot{V}_E(t) < 0$, it means that voltages of the DGs converge to their references, i.e., E^*, asymptotically. When $\dot{V}_E(t) = 0$ then we get $e_k^E(t) = 0$ and $e_c^E(t) = e_d^E(t-\tau)$.

6.7 REACTIVE POWER-SHARING CONTROLLER

A reactive power-sharing controller considering CTDs can be presented as:

$$E_c = E^* - n_c Q_c + v_c^Q \quad (6.32)$$

$$\frac{dv_c^Q}{dt} = Q_Q \left[\sum_{d=1}^{n} b_{cd} \left[\left(Q_d(t-\tau) \right) - \left(Q_c(t) \right) \right] \right] \quad (6.33)$$

where Q_Q is the designed SC gain and v_c^Q is the SC variable. Moreover, $Q_c = Q_c^M / Q_c^*$ and $Q_d = Q_d^M / Q_d^*$, where Q_c^M and Q_d^M are the measured reactive power of c_{th}-DG and d_{th}-DG, respectively, while Q_c^* and Q_d^* are the rated reactive power of c_{th}-DG and d_{th}-DG, respectively. For simplicity, the dynamics in (6.33) can be presented in vector forms as:

$$\frac{dv_c^Q}{dt} = \mathbb{Z}_Q v(t) + \mathbb{Z}_{Q-1} v(t - \tau(t)) \quad (6.34)$$

where $\mathbb{Z}_Q$ and $\mathbb{Z}_{Q-1}$ are $n \times n$ matrices whose elements are positive real numbers. To derive the stability conditions, a Razumikhin's method is used by applying a Lyapunov function as:

$$V_Q(v_t) = v^T(t) K_Q v(t) \quad (6.35)$$

where K_Q is $n \times n$ matrix and $K_Q > 0$, it should be noted that K_Q is symmetric and positive definite. Consider a time derivative of $V_Q(v_t)$ along (6.34) and apply the Lyapunov Razumikhin's method as $\bar{\varsigma}_P(s) = \bar{\varsigma}_Q \cdot s$, where $\bar{\varsigma}_Q > 0$. Hence, according to Razumikhin's theorem when the condition

$$\bar{\varsigma}_Q v^T(t) K_Q v(t) - v^T(t - \tau(t)) K_Q v(t - \tau(t)) \geq 0 \quad (6.36)$$

holds for some constant such that $\bar{\zeta}_Q = 1 + \zeta_Q$, where $\zeta_Q > 0$. Thus, it can be said that for any $\kappa_Q > 0$ there exists $\gamma_Q > 0$ such that:

$$\dot{V}_Q\big(v(t)\big) = 2v^T(t)K_Q\Big[\mathbb{Z}_Q v(t) + \mathbb{Z}_{Q-1} v\big(t - \tau(t)\big)\Big]$$

$$\leq 2v^T(t)K_Q\Big[\mathbb{Z}_Q v(t) + \mathbb{Z}_{Q-1} v\big(t - \tau(t)\big)\Big] \tag{6.37}$$

$$+ \kappa_Q\Big[\bar{\zeta}_Q v^T(t)K_Q v(t) - v^T\big(t - \tau(t)\big)K_Q v\big(t - \tau(t)\big)\Big] \leq -\gamma_Q |v(t)|^2$$

If

$$\begin{bmatrix} \mathbb{Z}_Q^T K_Q + K_Q \mathbb{Z}_Q + \kappa_Q K_Q & K_Q \mathbb{Z}_{Q-1} \\ \mathbb{Z}_{Q-1}^T K_Q & -\kappa_Q K_Q \end{bmatrix} < 0 \tag{6.38}$$

From (6.37) and (6.38), it can be observed that the matrix inequality is independent of the delay derivative term; hence, it can be concluded that there is no constraint on delay derivative. Hence, (6.37) provides sufficient conditions for the asymptotically and independent stability of the system having CTDs.

Remark 6.2

To regulate the frequency and voltage while ensuring appropriate active and reactive power sharing, the secondary control laws, i.e., ϖ_c^ω in (6.7), ϖ_c^P in (6.17), v_c^E in (6.24), and v_c^Q in (6.33) are designed, respectively. These secondary control laws ensure the system's stability in the presence of CTDs.

A schematic of proposed distributed controller is presented in Figure 6.2.

6.8 DISCUSSION AND PERFORMANCE ANALYSIS

To validate the efficacy and performance of the delay-independent distributed SC, simulations are conducted in a MATLAB/SimPower environment. An MG system that is considered in this work consists of four radial buses, four-DGs (DG_1-DG_4), four local loads (L_1-L_4), and are connected through a physical network whose line impedances are denoted by (Z) as shown in Figure 6.3. The values of different parameters and ratings used in this simulation study are presented in Table 6.2. Different scenarios and cases are considered to validate the performance of the presented controller which are discussed below in detail Ref. [26].

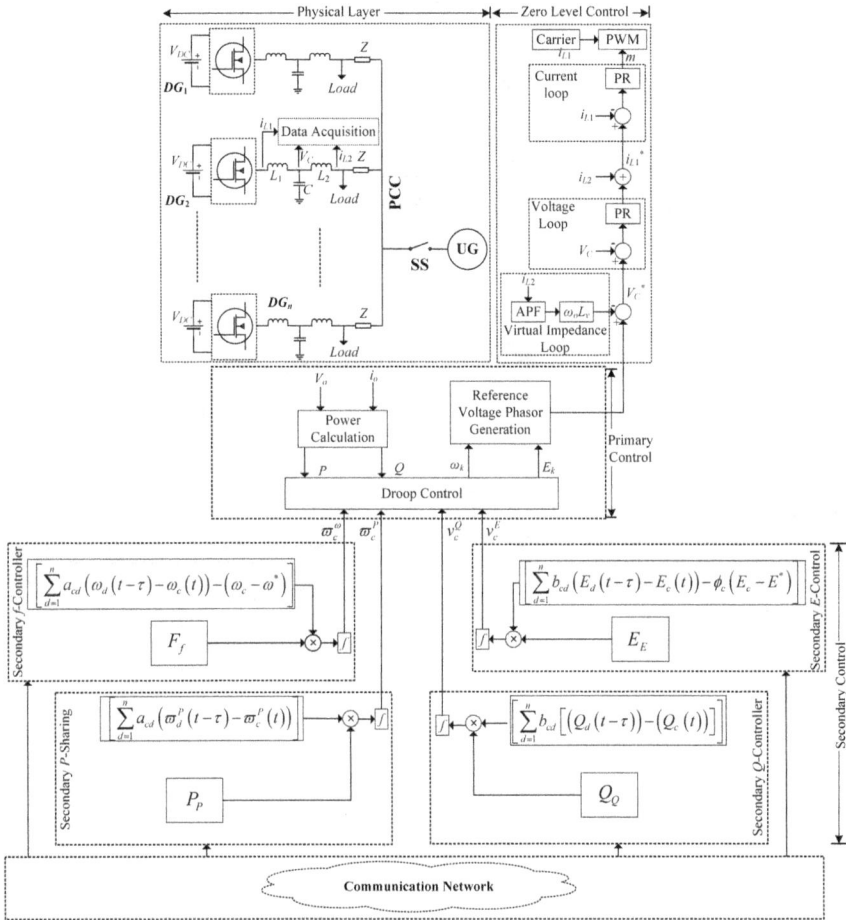

FIGURE 6.2   Schematic of distributed control structure.

6.8.1 Comparison Study

As discussed above, most controllers are delay-dependent, thus requiring a buffer or compensator to mitigate the effect of CTDs [7,9,13,14,18,19]. Moreover, it is also discussed that the proposed controller is delay-independent and does not require any compensator for CTDs' mitigation. Hence, to validate this point and show the advantageous features and robustness of the proposed controller over the other state-of-the-art controllers, a comparative study is performed. Specifically, the proposed controller is compared with the finite time control scheme designed for CTD mitigation as proposed in Ref. [9].

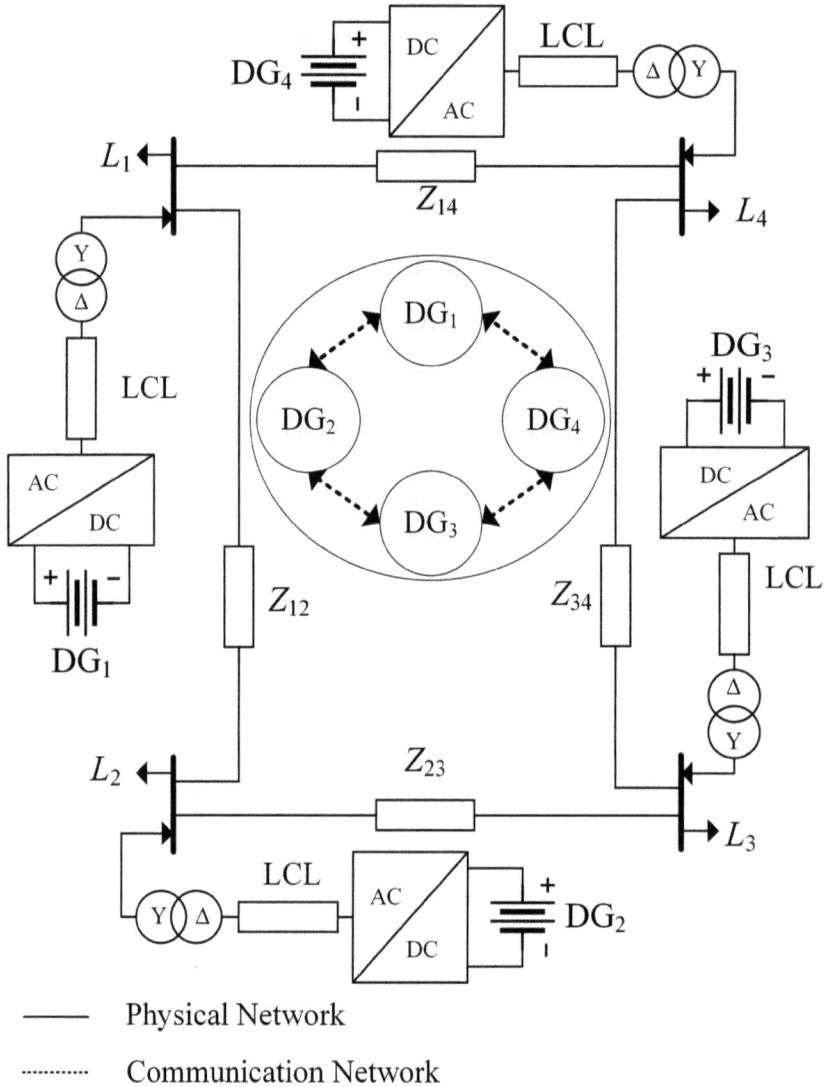

FIGURE 6.3 MG system under consideration.

TABLE 6.2 Parameters Used in this Study

Parameter			Symbol	Value
		Electrical		
Frequency	Switching		f_{sw}	10×10^{-3} Hz
	Nominal		f	50 Hz
	Sampling		f_a	$1/1 \times 10^{-6}$ Hz
				(*Continued*)

TABLE 6.2 (*Continued*)   Parameters Used in this Study

Parameter		Symbol	Value
	Electrical		
Nominal Voltage		E	310 V
Line Impedance		Z_{01}, Z_{14}	0.8 Ω + 3 mH
		Z_{12}	1.6 Ω + 6 mH
		Z_{23}	0.9 Ω + 4 mH
	Public	L_0	2 kW + 3 kVar
Loads	Local	$L_1 = L_2 = L_3 = L_4$	1 kW + 1 kVar
	LCL Filter		
Capacitance		C_f	3.31×10^{-5} F
Inductance	Inverter side	L_i	1.74×10^{-4} H
	Grid side	l_g	1.2×10^{-3} H
	Droop Controller		
Droop coefficients	P–ω	m_c	10^{-5} rad / (W s)
	Q–E	n_c	10^{-3} V/Var
	Secondary Controller		
SC gains	Frequency gain	F_f	0.1
	Voltage gain	E_E	0.01
	Active power gain	P_p	0.021
	Reactive power gain	Q_Q	0.013

Initially, the performance of the controller presented in Ref. [9] is checked by considering the CTDs equal to 0.2 sec, which are introduced at $t=2$ sec. It should be noted that in this scenario, a controller of Ref. [9] is considered without a compensator, so no additional buffer is used to compensate for the effect of CTDs. The simulation results in this case are shown in Figure 6.4. From the results, it can be seen that at $t=2$ sec, when CTDs are introduced in the system, the controller in Ref. [9] cannot handle the disturbances and lead the system towards instability. Hence, a controller with a compensator or buffer to mitigate the effects of CTDs is not a feasible solution as it requires electrical power for its operation. Moreover, it requires high computation-processor large memory storage, and also results in control complexity [26]. To avoid the use of a buffer or compensator, this chapter presents a delay-independent buffer-free distributed controller to mitigate the CTDs effects. The simulation results of the proposed controller under the same conditions are shown in Figure 6.5. From simulation results, it can be seen that at $t=2$ sec, when CTDs are imposed in the CN, some deviations in waveforms from their references are noticed but are restored very efficiently.

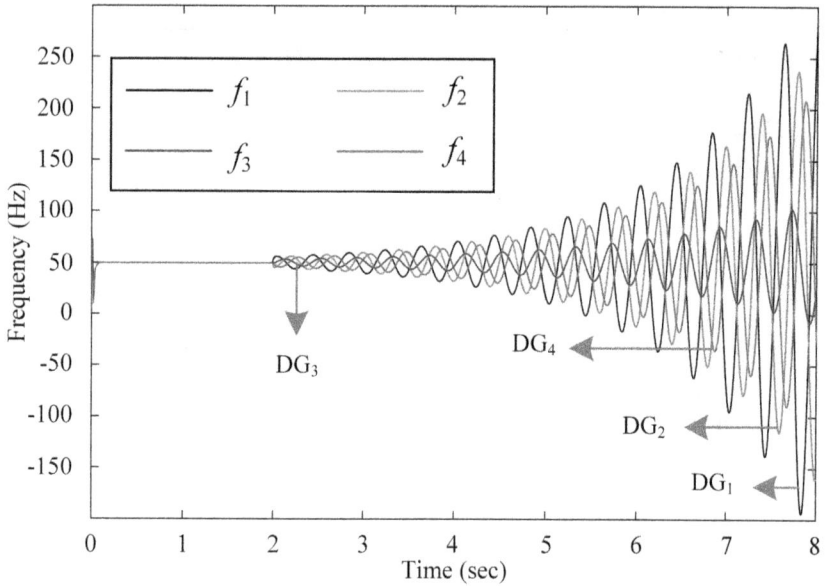

FIGURE 6.4a Frequency waveforms of DGs when controller in Ref. [9] is used without buffer compensator and considering $\tau=0.2$ sec.

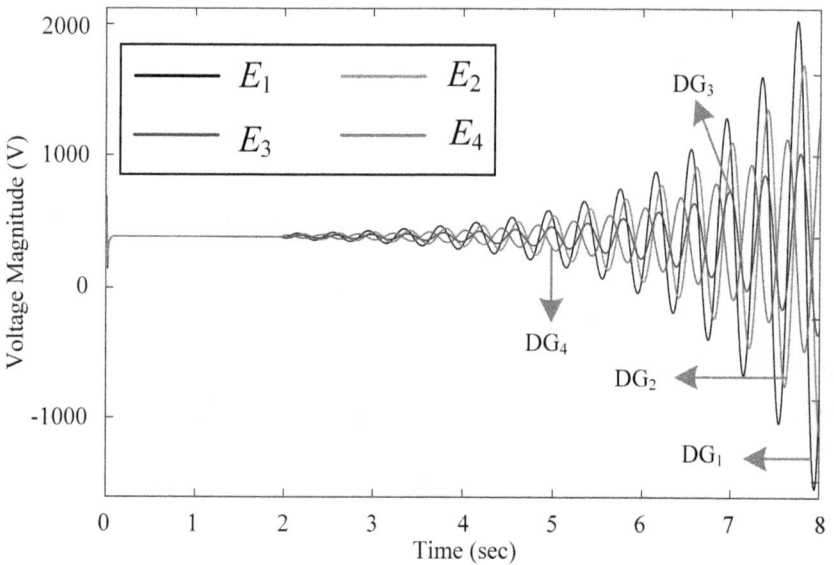

FIGURE 6.4b Voltage waveforms of DGs when controller in Ref. [9] is used without buffer compensator and considering $\tau=0.2$ sec.

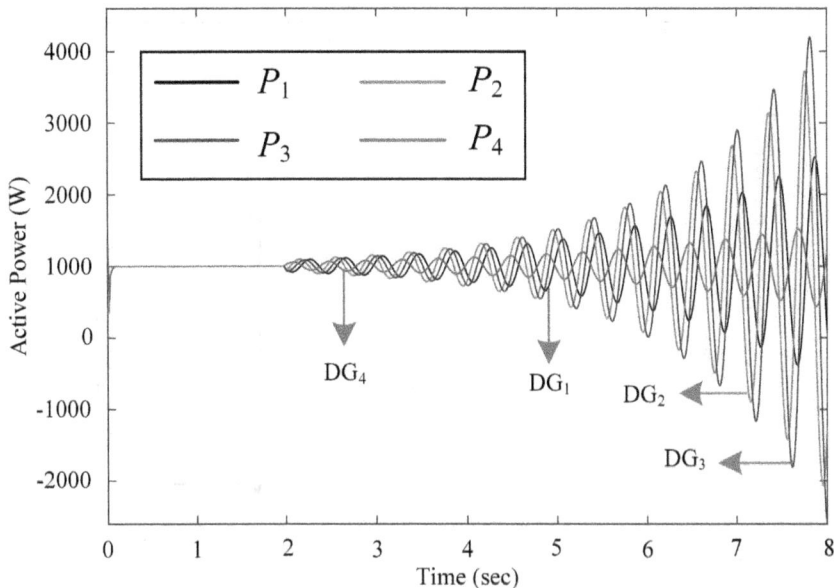

FIGURE 6.4c   Active power waveforms of DGs when controller in Ref. [9] is used without buffer compensator and considering $\tau=0.2$ sec.

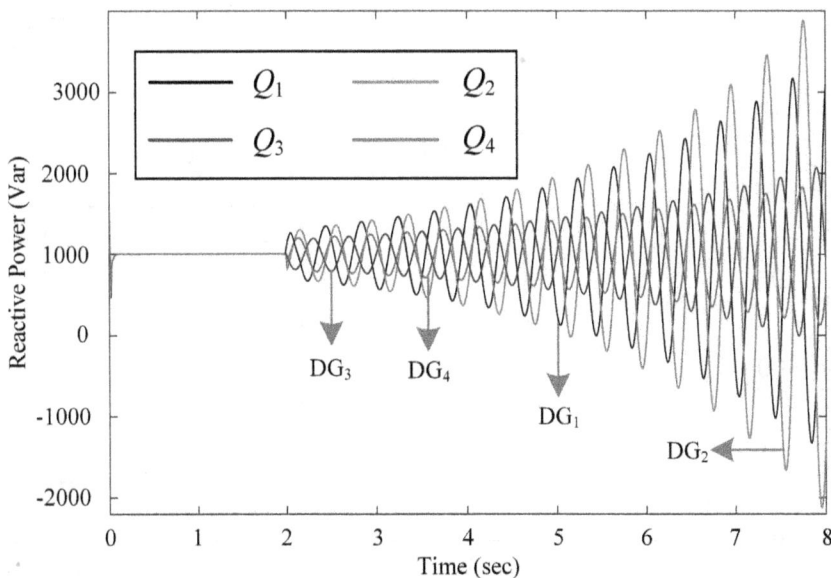

FIGURE 6.4d   Reactive power waveforms of DGs when controller in Ref. [9] is used without buffer compensator and considering $\tau=0.2$ sec.

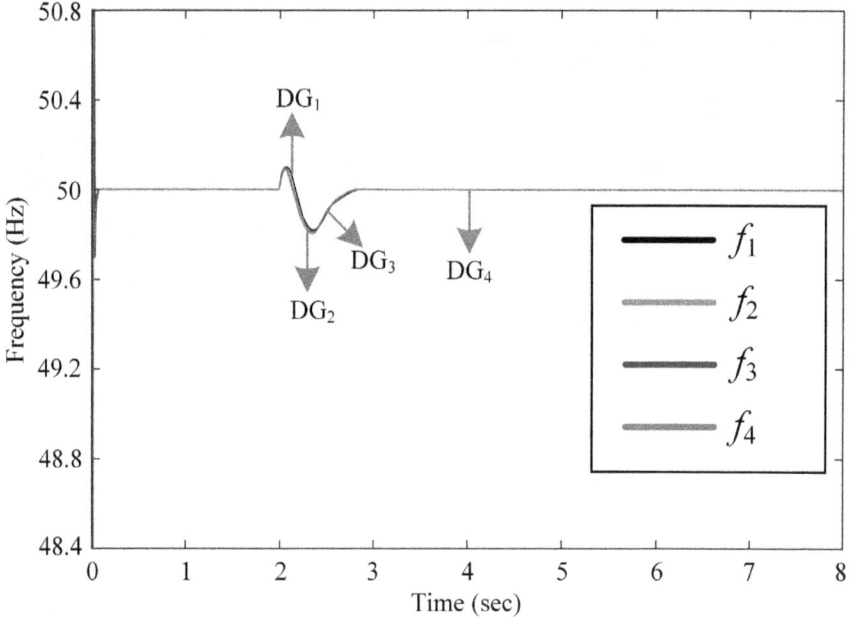

FIGURE 6.5a   Frequency waveforms of DGs considering $\tau = 0.2$ sec.

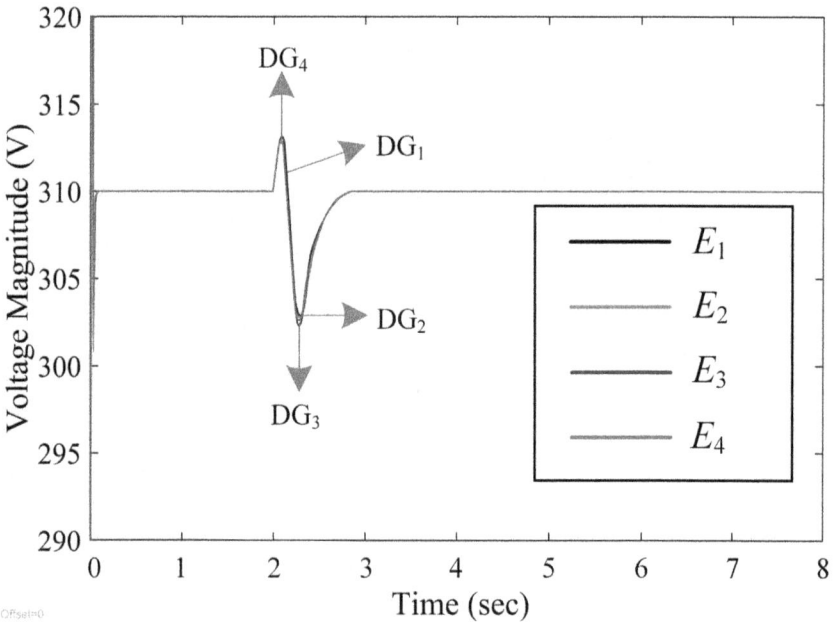

FIGURE 6.5b   Voltage waveforms considering $\tau = 0.2$ sec.

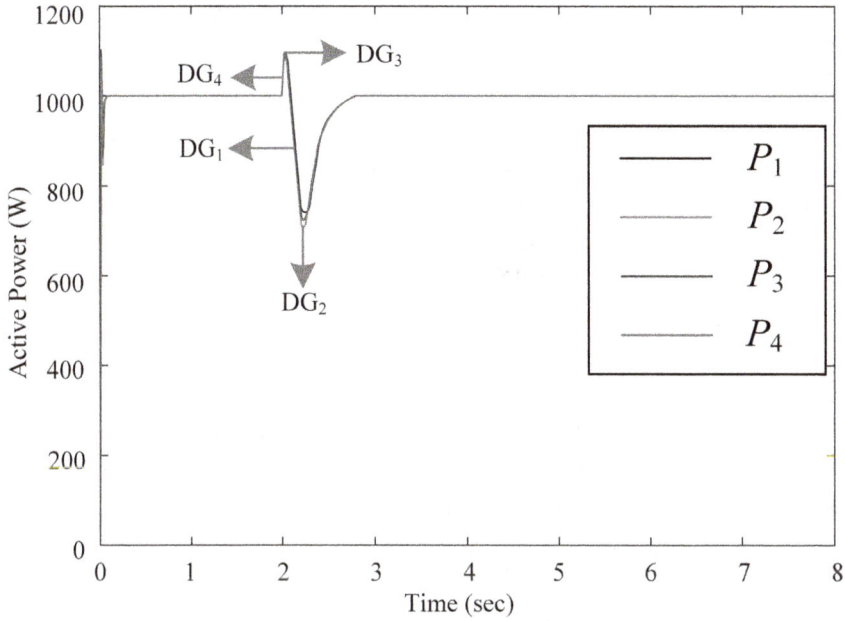

FIGURE 6.5c   Active power waveforms considering $\tau=0.2$ sec.

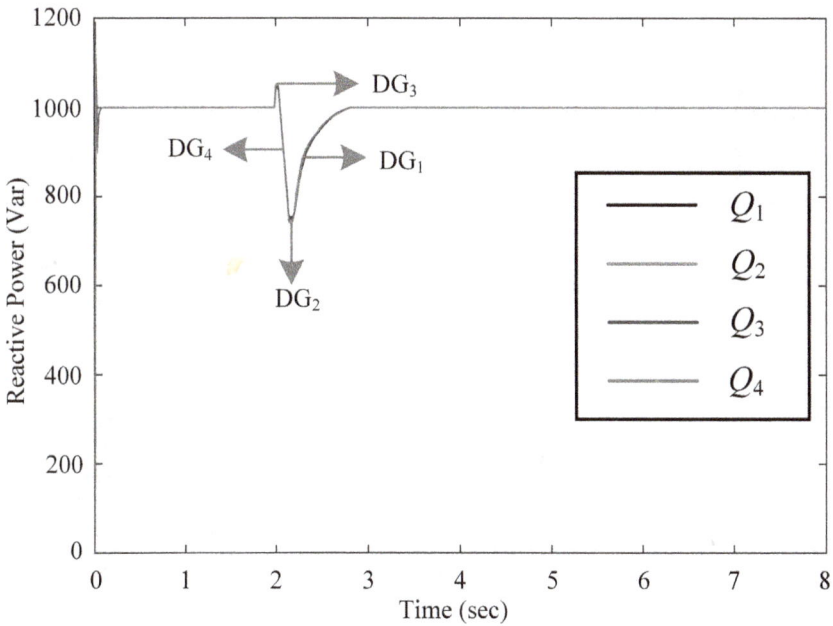

FIGURE 6.5d   Reactive power waveforms considering $\tau=0.2$ sec.

6.8.2 Performance of Controller Considering Different CTDs

The effectiveness of the controller is validated in a scenario in which it is assumed that all the cyber-links between the DGs are facing different CTDs. Hence, at $t=2$ sec, a set of CTDs, such that $\tau_1=0.05$ sec, $\tau_2=0.3$ sec, $\tau_3=0.5$ sec, and $\tau_4=0.7$ sec, are introduced among the DGs. Upon the injection of different CTDs at $t=2$ sec, a deviation in the output waveforms of an MG is observed, as shown in Figure 6.6. From Figure 6.6, it can be seen that in case of small CTD, that is, $\tau_1=0.05$ sec, which is introduced in cyber-link between DG_1 and DG_2, the proposed controller performs efficiently, and within around 0.55 sec, the deviations are restored. However, in case of $\tau_4=0.7$ sec, a large deviation in output waveforms is observed but is restored by the controller within 1.7 sec. This is because that the proposed controller is delay-independent therefore it is capable of performing effectively even if the delays are large.

6.8.3 Plug-and-Play Functionality

As the energy generation from the RESs is stochastic and intermittent in nature, therefore, an MG must be able to allow a DG to be attached or detached at any instant without redesigning its existing structure. Hence,

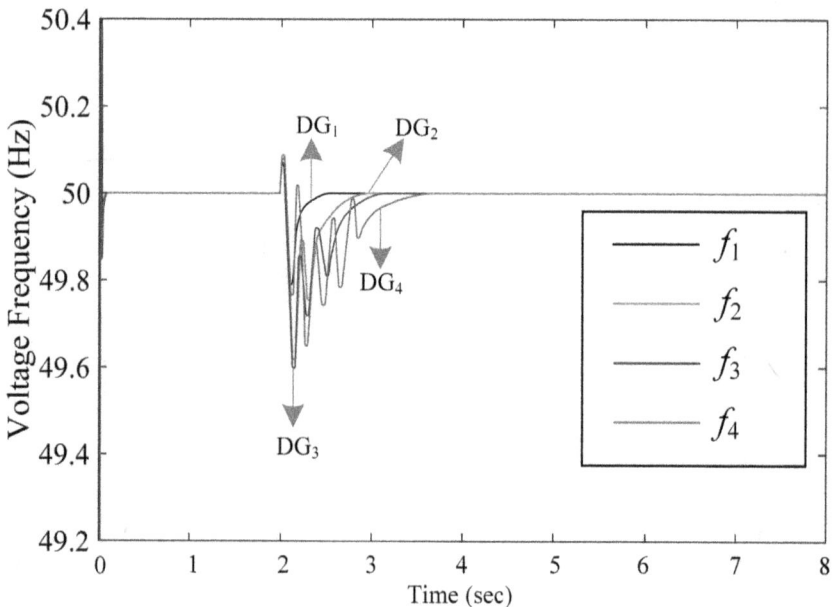

FIGURE 6.6a   Frequency waveforms of DGs considering different CTDs.

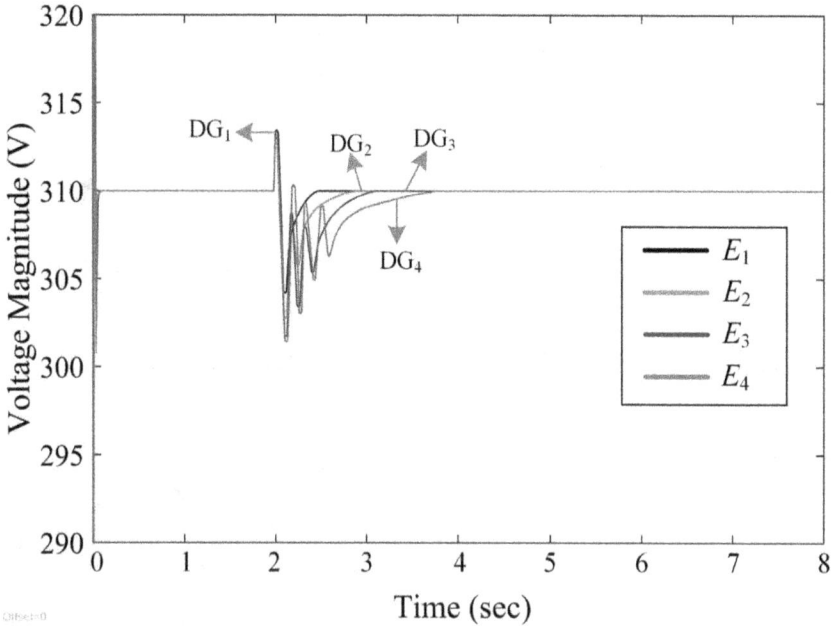

FIGURE 6.6b   Voltage waveforms of DGs considering different CTDs.

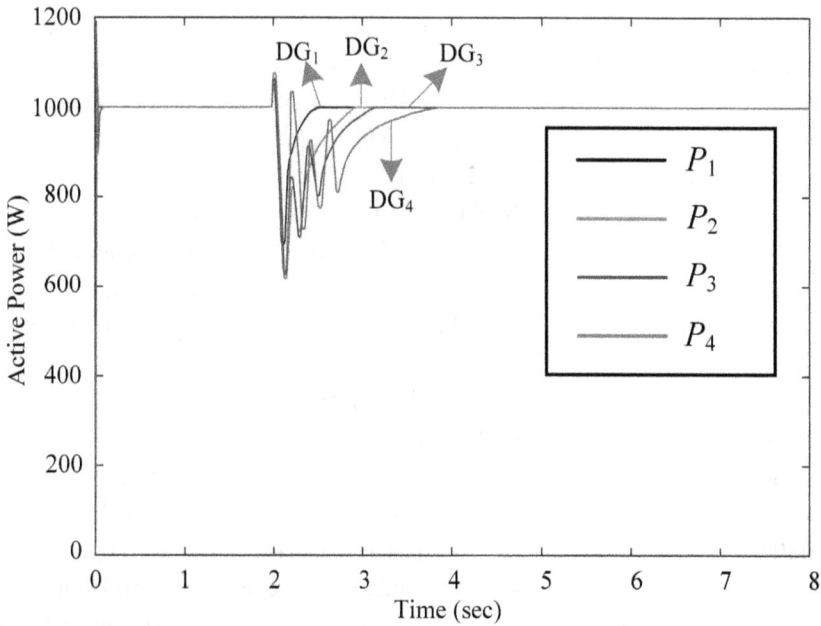

FIGURE 6.6c   Active power waveforms of DGs considering different CTDs.

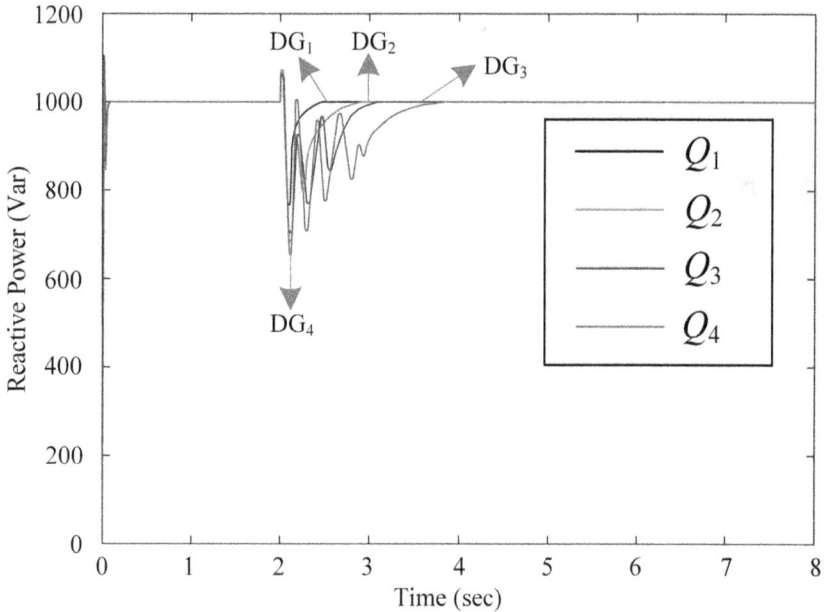

FIGURE 6.6d   Reactive power waveforms of DGs considering different CTDs.

it is very important for the stable and smooth operation of an MG that the controller must be able to regulate the frequency and voltage and ensure accurate sharing of load among the DGs if a system is under such conditions. To validate the controller's performance during PnP operation, it is assumed that from $0 \leq t \leq 2$, an MG performs its normal operation; however, at $t=2$ sec, a DG_5 is plugged-in into the system and plugged-out of the system at $t=4$ sec (cyber-physical networks are also connected and disconnected). During PnP functionality, when DGs are plugged-in or plugged-out the physical and cyber networks of the MG system change accordingly, as shown in Figures 6.7 and 6.8, respectively. To validate the controller's performance during PnP functionality, simulation tests are performed by considering small as well as large CTDs.

The simulation results of the presented control scheme during PnP operation while considering that all the communication links are subjected to small CTDs, that is, $\tau=0.05$ sec, are shown in Figure 6.9. In a scenario, where DG_5 is plugged-in into the system at $t=2$ sec, some deviations in frequency and voltage waveforms of the DGs are observed, but the controller restores these waveforms very rapidly, and within about 0.4 sec, the waveforms come to their steady-state condition. Moreover, a maximum deviation of about

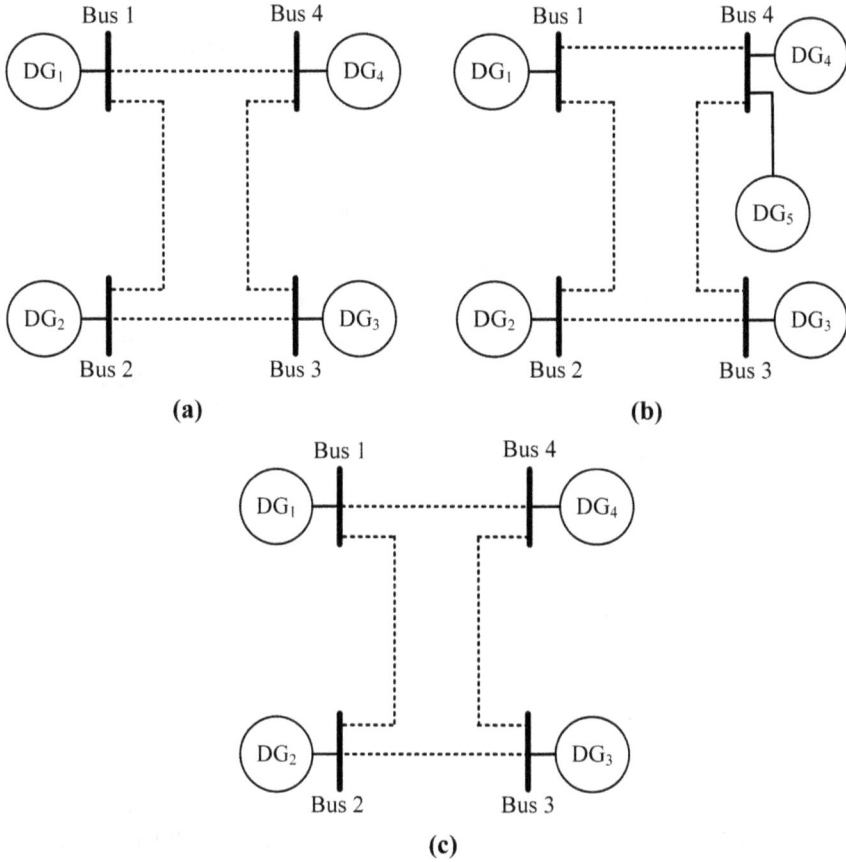

FIGURE 6.7 MG physical network topology during (a) normal operation (b) DG_5 plug-in (c) DG_5 plug-out.

0.25 Hz and 5 V in the output frequency and voltage waveforms of DGs is observed, as shown in Figure 6.9a and b, respectively. Similarly, at the same time, a new power consensus is generated, and the load is proportionally shared among all the DGs, as can be seen in Figure 6.9c and d. Moreover, when DG_5 is plugged-out of the MG system, that is, at $t=4$ sec, some deviations in frequency and voltage waveforms are noticed but are restored by the controller with high efficacy as presented in Figure 6.9. Moreover, from simulation results, it can be observed that during the plug-in or plug-out operation, a new power consensus is generated in both scenarios, and all the load is proportionally shared among the DGs.

When the communication links among the DGs are subjected to large CTDs, that is, $\tau=0.5$ sec, during PnP operation, the performance of the

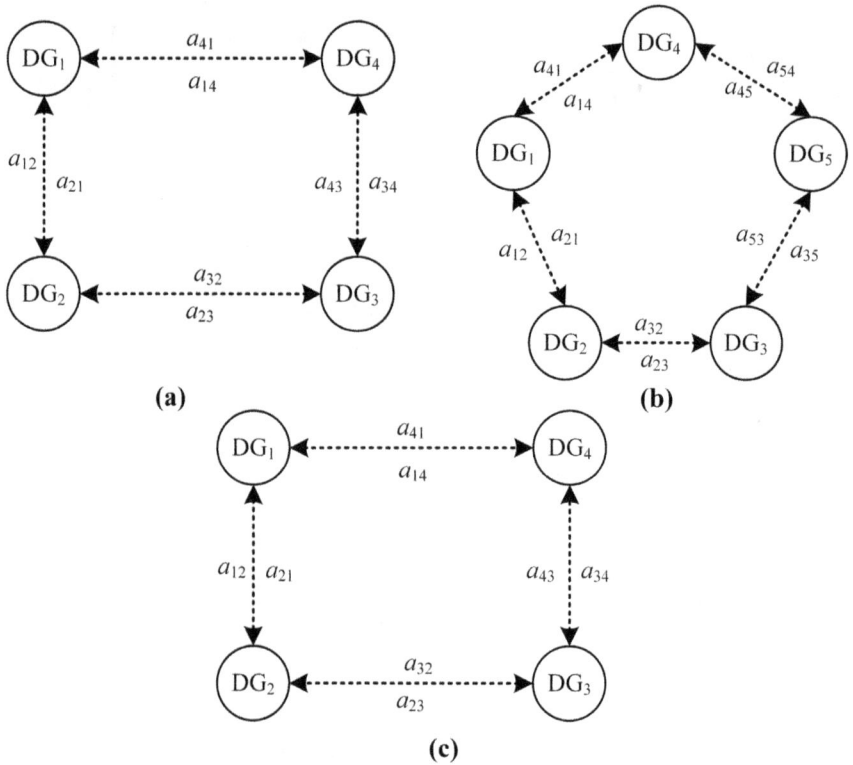

FIGURE 6.8 MG CN topology during (a) normal operation (b) DG_5 plug-in (c) DG_5 plug-out.

controller in this scenario is presented in Figure 6.10. The output frequency and voltage waveforms of DGs during the plug-in and plug-out phenomena are presented in Figure 6.10a and b. From Figure 6.10a and b, a deviation of waveforms from their references is observed, but the controller restores the deviation with a fast dynamic response, and within 0.9 sec they come to their steady-state, having a maximum deviation of about 0.3 Hz in frequency and 7.4 V in voltage. Similarly, when DG_5 is plugged-out of the MG system at $t=4$ sec, the same steady-state and dynamic responses are observed. Furthermore, a new power consensus is generated at $t=2$ sec and $t=4$ sec, and the loads are accurately distributed among all the DGs, as can be seen in Figure 6.10c and d. From the simulation results, it can be seen that the proposed controller performs effectively during PnP operation even in the presence of large CTDs.

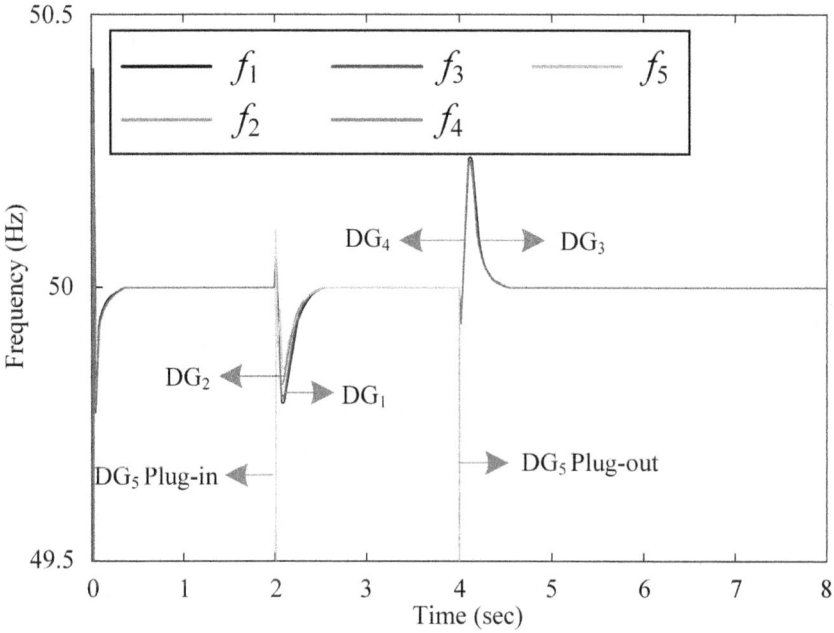

FIGURE 6.9a   Frequency waveforms of DGs under PnP considering $\tau = 0.05$ sec.

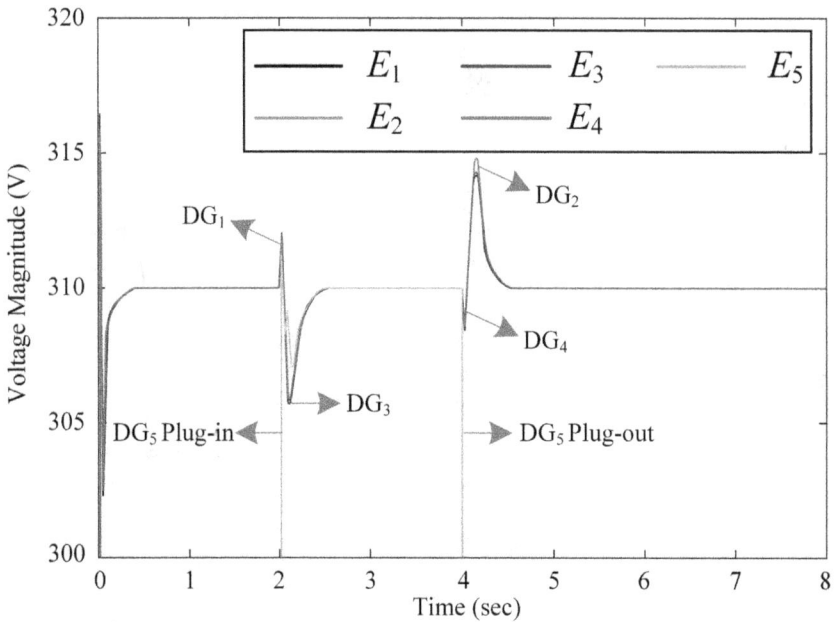

FIGURE 6.9b   Voltage waveforms of DGs under PnP considering $\tau = 0.05$ sec.

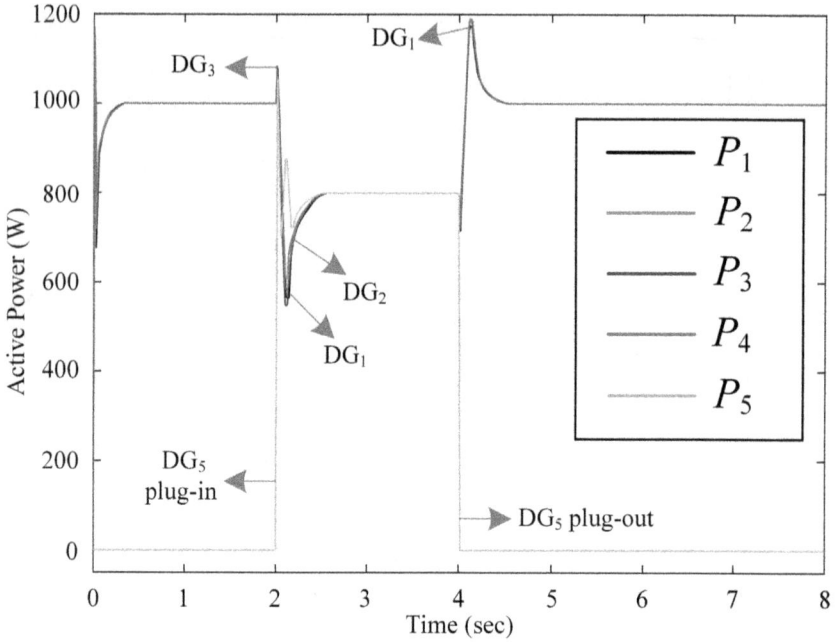

FIGURE 6.9c   Active power waveforms of DGs under PnP considering $\tau=0.05$ sec.

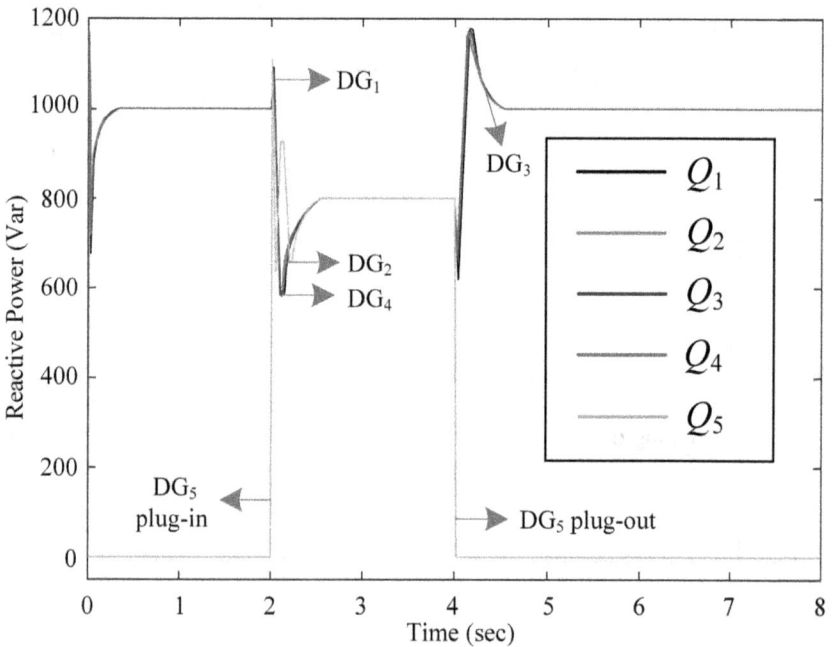

FIGURE 6.9d   Reactive waveforms of DGs under PnP considering $\tau=0.05$ sec.

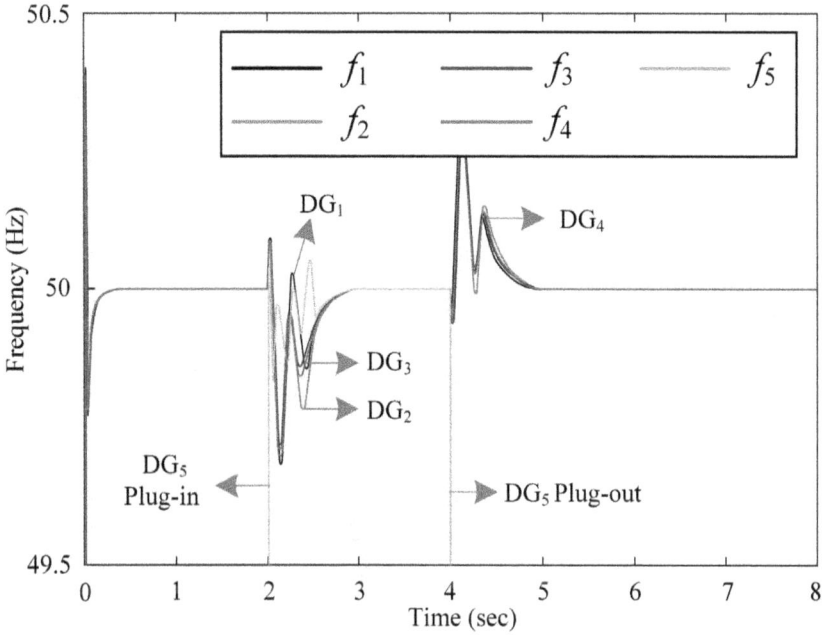

FIGURE 6.10a   Frequency waveforms of DGs under PnP considering $\tau = 0.5$ sec.

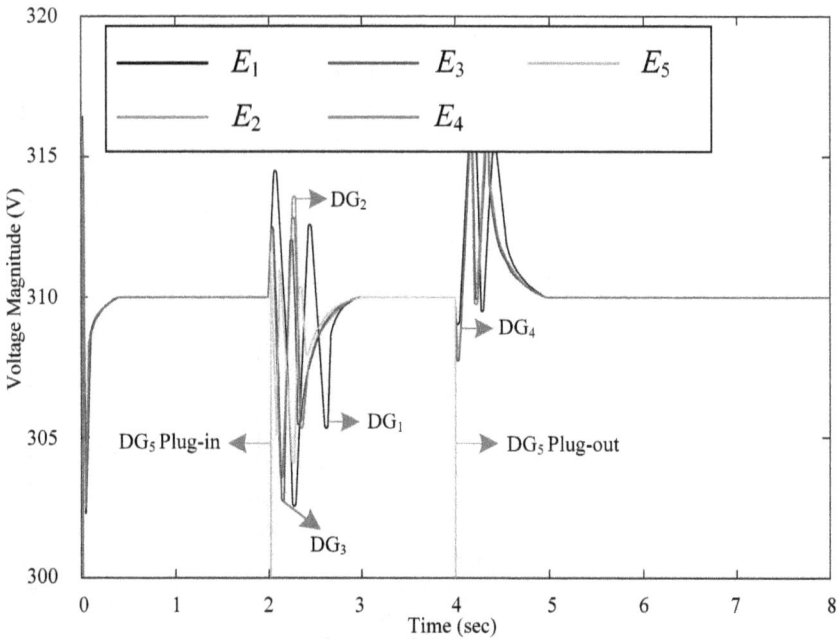

FIGURE 6.10b   Voltage waveforms of DGs under PnP considering $\tau = 0.5$ sec.

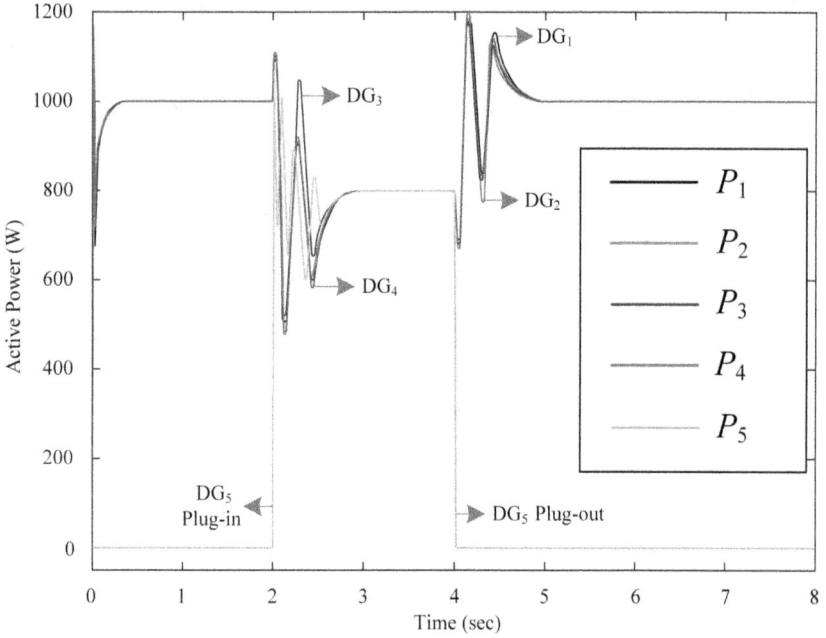

FIGURE 6.10c   Active power waveforms of DGs under PnP considering $\tau = 0.5$ sec.

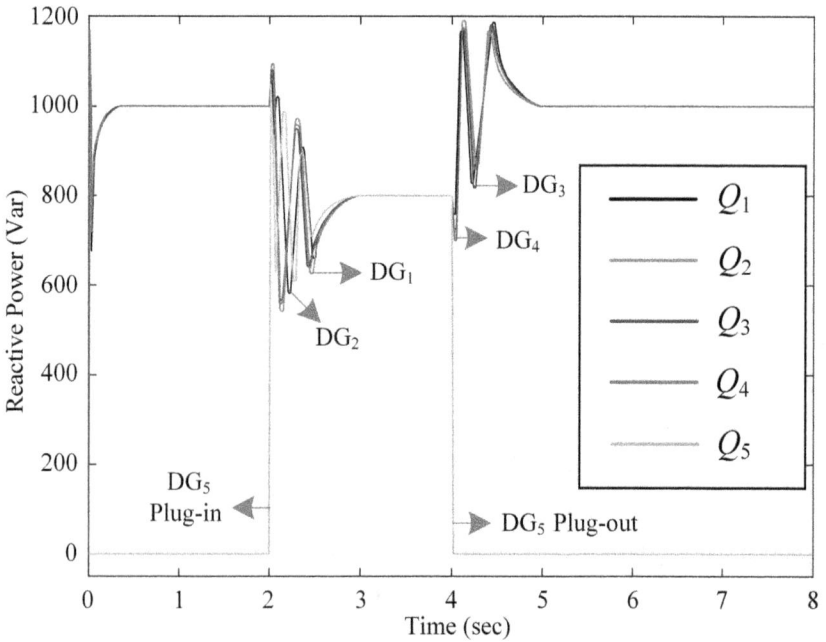

FIGURE 6.10d   Reactive waveforms of DGs under PnP considering $\tau = 0.5$ sec.

TABLE 6.3   Variation in Load with Respect to Time

Time Period (sec)	$0 \leq t \leq 2$	$2 \leq t \leq 4$	$4 \leq t \leq 8$
L_1, L_2	1+1	2+2	1+1
L_3, L_4	1+1	0.5+0.5	1+1

6.8.4 Load Variation Condition

Based on consumer activity, a load changes can occur at any instant in an MG; therefore, the controller must be robust enough to ensure the system's stability in case of any load variation. Hence, to validate the controllers' performance, it is assumed that the load varies according to the pattern shown in Table 6.3. From Table 6.3, it can be seen that during interval $0 \leq t \leq 2$, the values of loads are $L_1 = L_2 = L_3 = L_4 = 1$ kW + 1 kVar. At $t=2$ sec, the loads change during interval $2 \leq t \leq 4$, the values of loads become $L_1 = L_2 = 2$ kW + 2 kVar and $L_3 = L_4 = 0.5$ kW + 0.5 kVar. At $t=4$ sec, the load varies again and during the time interval the values of the load become $L_1 = L_2 = L_3 = L_4 = 1$ kW + 1 kVar.

When the load varies with time according to the pattern shown in Table 6.3, while considering CTDs ($\tau = 0.05$ sec), the performance of the controller can be seen in Figure 6.11. From the simulation results, it can be seen that when a load variation occurs, the controller restores the frequencies and voltages of DGs effectively, as presented in Figure 6.11a and b, respectively. Although some deviations from their references occur, i.e., in case of frequency, a maximum deviation of 0.2 Hz occurs while in case of voltage a maximum deviation of 3.5 V is observed, the proposed controller restores these deviations very efficiently, and within 0.55 sec, these deviations come to their steady-state and accurately track their references. Moreover, at $t=2$ sec, when the load changes from 4 kW + 4 kVar to 5 kW + 5 kVar, a new and updated power consensus is created, and the load is distributed homogeneously between all DGs as presented in Figure 6.11c and d. Similarly, at $t=4$ sec, when a variation in load occurs again, the same dynamic and steady-state responses from the controller are observed, as can be seen in Figure 6.11.

Similarly, to validate the effectiveness of the controller in case of load variations while considering the CTDs ($\tau = 0.5$ sec) among the cyber-links, simulations are performed as presented in Figure 6.12. The frequencies and voltages of the DGs are presented in Figure 6.12a and b, respectively. From Figure 6.12a and b, it can be noticed that when the load changes according to the pattern presented in Table 6.3, some deviations are observed, but the controller restores these deviations in 1.1 sec with a maximum deviation of 0.4 Hz and 8.5 V in frequency and voltage, respectively. The active

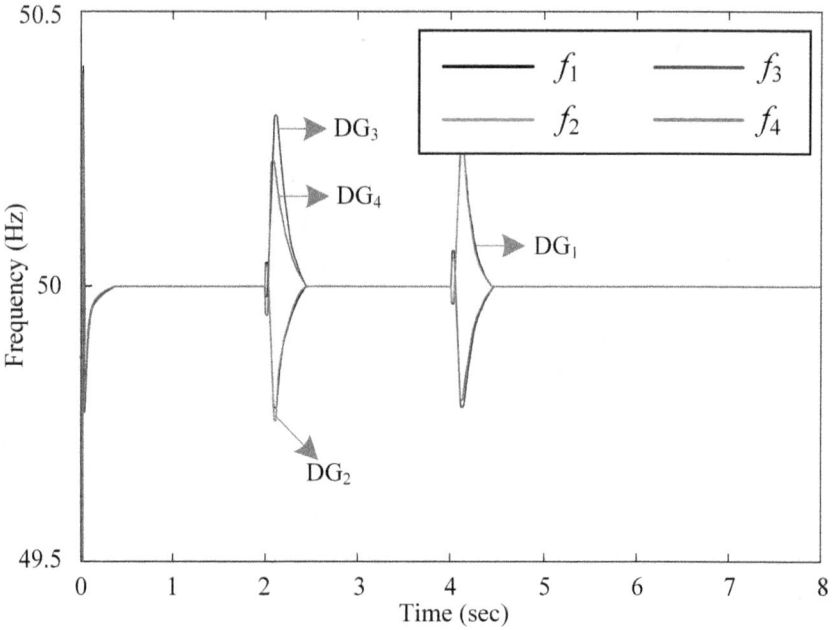

FIGURE 6.11a Frequency waveforms of DGs under load variation considering $\tau = 0.05$ sec.

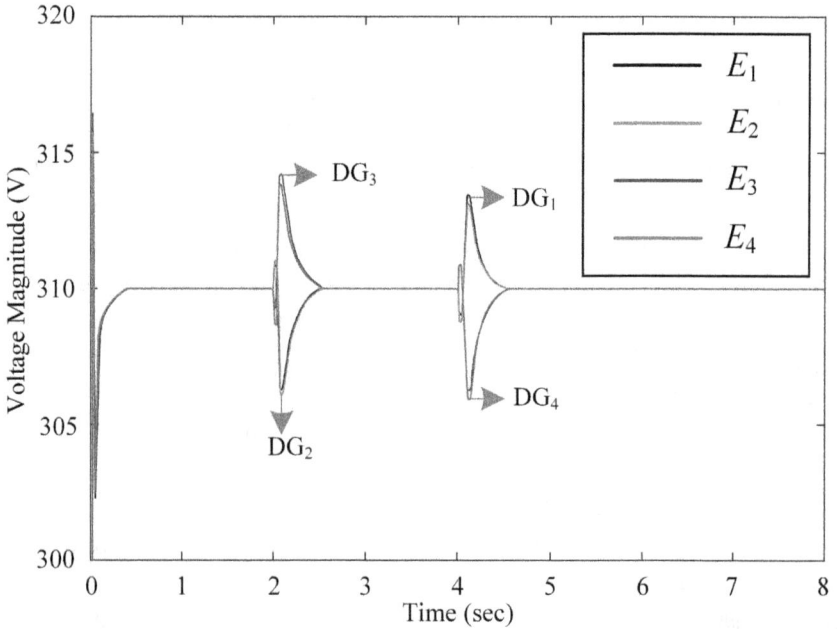

FIGURE 6.11b Voltage waveforms of DGs under load variation considering $\tau = 0.05$ sec.

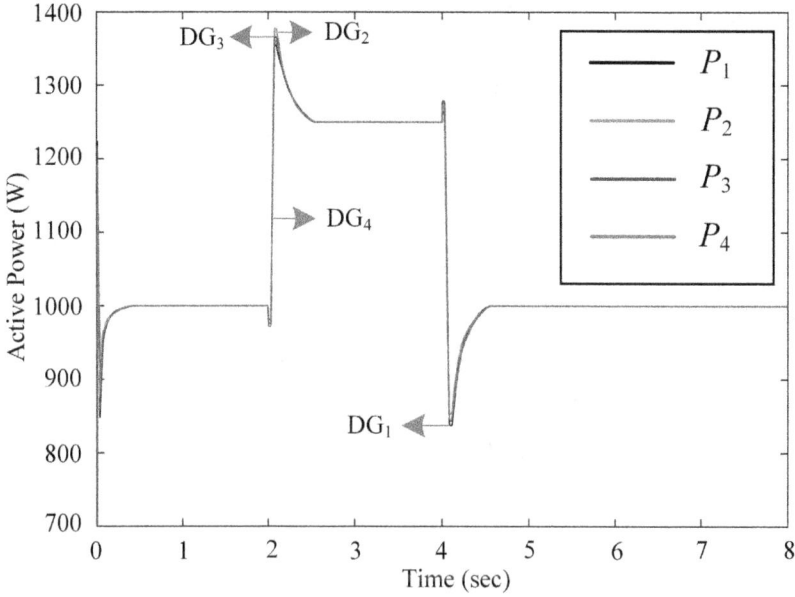

FIGURE 6.11c   Active power waveforms of DGs under load variation considering $\tau = 0.05$ sec.

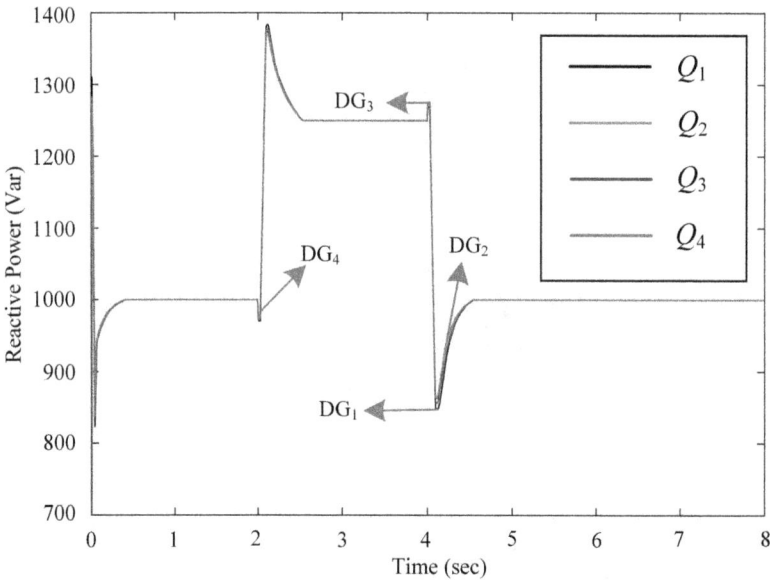

FIGURE 6.11d   Reactive power waveforms of DGs under load variation considering $\tau = 0.05$ sec.

and reactive power waveforms are shown in Figure 6.12c and d, respectively. From these output waveforms, it can be noticed that accurate power sharing is achieved in load variation conditions, even if large CTDs are considered.

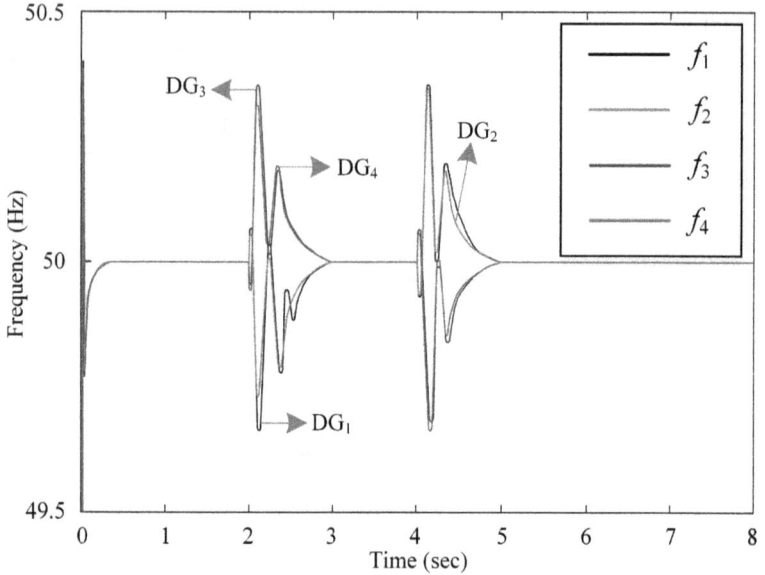

FIGURE 6.12a   Frequency waveforms of DGs under load variation considering $\tau=0.5$ sec.

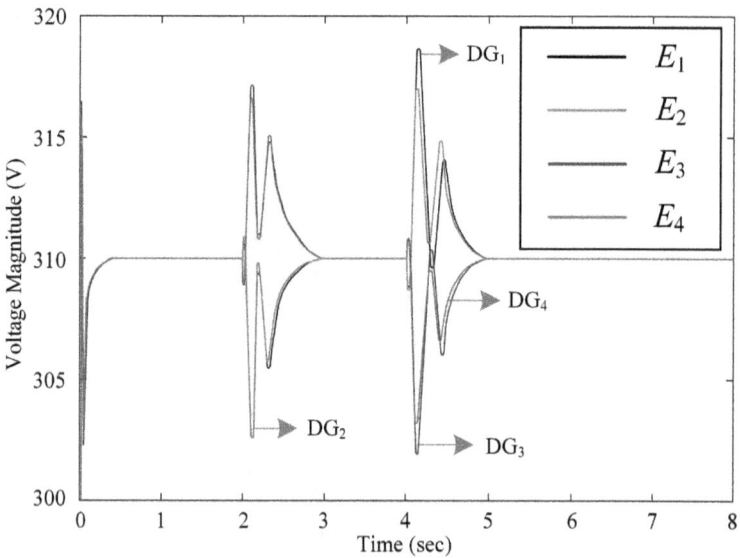

FIGURE 6.12b   Voltage waveforms of DGs under load variation considering $\tau=0.5$ sec.

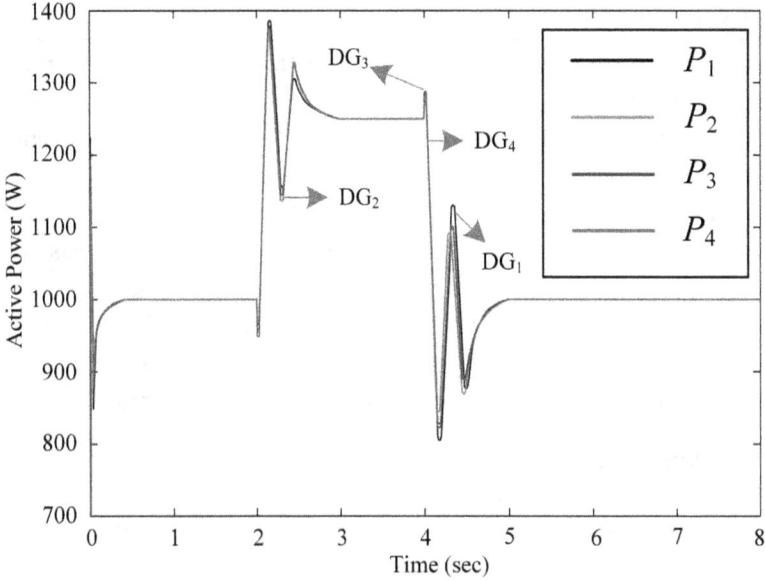

FIGURE 6.12c   Active power waveforms of DGs under load variation considering $\tau = 0.5$ sec.

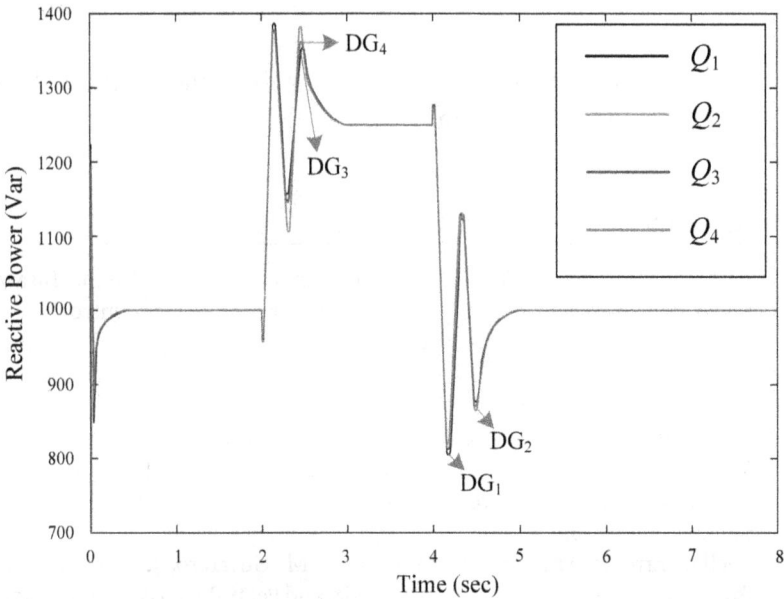

FIGURE 6.12d   Reactive power waveforms of DGs under load variation considering $\tau = 0.5$ sec.

6.9 CONCLUSION

In this chapter, a delay-independent buffer-free distributed controller is presented for the MG system that can regulate the voltage and frequency and ensure appropriate load sharing among the DGs in the presence of CTDs. The presented controller can mitigate small as well as large, uniform and non-uniform CTDs. Moreover, the stability of the system is ensured by using a Lyapunov function, and a sensitivity analysis is also provided that represents the system's stability and provides a range of CTDs. Furthermore, the controller's performance is validated through simulation results that are performed in different scenarios, such as PnP and load variations in the presence of uniform and non-uniform CTDs.

6.10 EXERCISES

1. Using the state-of-art communication network, what will be the maximum possible time delay of the proposed controller?

2. Simulate the delay-independent control strategies presented in Refs. [4,6,10] and compare the results with the controller presented in this chapter.

3. Consider a scenario where the communication delays exist during the frequency and voltage regulation while no communication delays exist in active and reactive power sharing. Solve the equations in this scenario and simulate them.

REFERENCES

1. Rustam, Muhammad Asfandyar, Muhammad Yasir Ali Khan, Tasawar Abbas, and Bilal Khan. "Distributed secondary frequency control scheme with A-symmetric time varying communication delays and switching topology." *e-Prime-Advances in Electrical Engineering, Electronics and Energy* 9 (2024): 100650.
2. Zhou, Quan, Mohammad Shahidehpour, Ahmed Alabdulwahab, Abdullah Abusorrah, Liang Che, and Xuan Liu. "Cross-layer distributed control strategy for cyber resilient microgrids." *IEEE Transactions on Smart Grid* 12, no. 5 (2021): 3705–3717.
3. Coelho, Ernane Antonio, Dan Wu, Josep M. Guerrero, Juan C. Vasquez, Tomislav Dragicević, Cedomir Stefanović, and Petar Popovski. "Small-signal analysis of the microgrid secondary control considering a communication time delay." *IEEE Transactions on Industrial Electronics* 63, no. 10 (2016): 6257–6269.

4. Liu, Shichao, Xiaoyu Wang, and Peter Xiaoping Liu. "Impact of communication delays on secondary frequency control in an islanded microgrid." *IEEE Transactions on Industrial Electronics* 62, no. 4 (2014): 2021–2031.

5. Shahab, Mohammad Ali, Babak Mozafari, Soodabeh Soleymani, Nima Mahdian Dehkordi, Hosein Mohammadnezhad Shourkaei, and Josep M. Guerrero. "Stochastic consensus-based control of μGs with communication delays and noises." *IEEE Transactions on Power Systems* 34, no. 5 (2019): 3573–3581.

6. Hashmi, Khurram, Muhammad Mansoor Khan, Muhammad Umair Shahid, Arshad Nawaz, Asad Khan, Jia Jun, and Houjun Tang. "An energy sharing scheme based on distributed average value estimations for islanded AC microgrids." *International Journal of Electrical Power & Energy Systems* 116 (2020): 105587.

7. Lai, Jingang, Hong Zhou, Xiaoqing Lu, Xinghuo Yu, and Wenshan Hu. "Droop-based distributed cooperative control for microgrids with time-varying delays." *IEEE Transactions on Smart Grid* 7, no. 4 (2016): 1775–1789.

8. Zhao, Binyan, Xiaodai Dong, and Jens Bornemann. "Service restoration for a renewable-powered microgrid in unscheduled island mode." *IEEE Transactions on Smart Grid* 6, no. 3 (2014): 1128–1136.

9. Ning, Boda, Qing-Long Han, and Lei Ding. "Distributed finite-time secondary frequency and voltage control for islanded microgrids with communication delays and switching topologies." *IEEE Transactions on Cybernetics* 51, no. 8 (2020): 3988–3999.

10. Yi, Zhongkai, Yinliang Xu, Wei Gu, and Zhongyang Fei. "Distributed model predictive control based secondary frequency regulation for a microgrid with massive distributed resources." *IEEE Transactions on Sustainable Energy* 12, no. 2 (2020): 1078–1089.

11. Shan, Yinghao, Jiefeng Hu, Ka Wing Chan, and Syed Islam. "A unified model predictive voltage and current control for microgrids with distributed fuzzy cooperative secondary control." *IEEE Transactions on Industrial Informatics* 17, no. 12 (2021): 8024–8034.

12. Wang, Fang, Bing Chen, Xiaoping Liu, and Chong Lin. "Finite-time adaptive fuzzy tracking control design for nonlinear systems." *IEEE Transactions on Fuzzy Systems* 26, no. 3 (2017): 1207–1216.

13. Raeispour, Mohammad, Hajar Atrianfar, Hamid Reza Baghaee, and Gevork B. Gharehpetian. "Resilient H∞ consensus-based control of autonomous AC microgrids with uncertain time-delayed communications." *IEEE Transactions on Smart Grid* 11, no. 5 (2020): 3871–3884.

14. Yao, Weitao, Yu Wang, Yan Xu, Chao Deng, and Qiuwei Wu. "Distributed weight-average-prediction control and stability analysis for an islanded microgrid with communication time delay." *IEEE Transactions on Power Systems* 37, no. 1 (2021): 330–342.

15. Yao, Weitao, Yu Wang, Yan Xu, and Chaoyu Dong. "Small-signal stability analysis and lead-lag compensation control for DC networked-microgrid under multiple time delays." *IEEE Transactions on Power Systems* 38, no. 1 (2022): 921–933.

16. Xie, Yijing, and Zongli Lin. "Distributed event-triggered secondary voltage control for microgrids with time delay." *IEEE Transactions on Systems, Man, and Cybernetics: Systems* 49, no. 8 (2019): 1582–1591.

17. Ullah, Shafaat, Laiq Khan, Irfan Sami, and Nasim Ullah. "Consensus-based delay-tolerant distributed secondary control strategy for droop controlled AC microgrids." *IEEE Access* 9 (2021): 6033–6049.

18. Hu, Wenqiang, Zaijun Wu, Xinxin Lv, and Venkata Dinavahi. "Robust secondary frequency control for virtual synchronous machine-based microgrid cluster using equivalent modeling." *IEEE Transactions on Smart Grid* 12, no. 4 (2021): 2879–2889.

19. Cai, Pengcheng, Chuanbo Wen, and Changju Song. "Consensus-based secondary frequency control for islanded microgrid with communication delays." In *2018 International Conference on Control, Automation and Information Sciences (ICCAIS)*, pp. 107–112. IEEE, Hangzhou, China, 2018.

20. Zhang, Wei, Wei Liu, Tian Wang, Anfeng Liu, Zhiwen Zeng, Houbing Song, and Shaobo Zhang. "Adaption resizing communication buffer to maximize lifetime and reduce delay for WVSNs." *IEEE Access* 7 (2019): 48266–48287.

21. Godsil, Chris, and Gordon F. Royle. *Algebraic Graph Theory*. Vol. 207. Cham, Switzerland: Springer Science & Business Media, 2001.

22. Schiffer, Johannes, Romeo Ortega, Alessandro Astolfi, Jörg Raisch, and Tevfik Sezi. "Conditions for stability of droop-controlled inverter-based microgrids." *Automatica* 50, no. 10 (2014): 2457–2469.

23. Kundur, Prabha. "Power system stability." *Power System Stability and Control* 10 (2007): 7–1.

24. Gu, Keqin, Jie Chen, and Vladimir L. Kharitonov. *Stability of Time-Delay Systems*. Cham, Switzerland: Springer Science & Business Media, 2003.

25. Fridman, Emilia. "Tutorial on Lyapunov-based methods for time-delay systems." *European Journal of Control* 20, no. 6 (2014): 271–283.

26. Huang, Xinghua, Yuanliang Fan, Han Wu, Gonglin Zhang, Zili Yin, Muhammad Yasir Ali Khan, Haoming Liu, and Jingjing Zhai. "Distributed secondary control for islanded microgrids considering communication delays." *IEEE Access* 12 (2024): 64335–64350.

Index

For Product Safety Concerns and Information please contact our EU
representative GPSR@taylorandfrancis.com
Taylor & Francis Verlag GmbH, Kaufingerstraße 24, 80331 München, Germany